PROBLEMS AND PERSPECTIVES OF WATERING THE CROPS

PROBLEMS AND PERSPECTIVES OF WATERING THE CROPS

(A Geographical Analysis of Rajasthan)

B.L. SHARMA

CONCEPT PUBLISHING COMPANY, NEW DELHI–110015

A Research project sponsored by the
University Grants Commission, New Delhi.

ISBN 81-7022-191-9

First Published 1987

Bhanwar Lal **Sharma** (b. 1938-

Published by
Naurang Rai
Concept Publishing Company
H-13, Bali Nagar
NEW DELHI-110015 (India)

Printed at
New Gian Offset Printers
495, Shahzada Bagh
DELHI-110035

Contents

List of Maps

List of Tables

ONE

Introduction

> "Irrigation is a package deal not a partial agreement. But in India, it is a partial agreement not a package deal."

Water is not merely a chemical composition of H_2O in scientific terminology, but in its natural condition, physical properties, types & quality and movements, etc. exhibit a complex relationship. Water is found in various aggregatory conditions and in different forms. It is found in gaseous, liquid and in moisture conditions which occur in different forms in different timings. Thus, it is a most important factor for development and growth of plants and fertility of soils. By translocating the products formed upon the interaction of organism and soil, it contributes to the formation of genetic soil horizons. From a meliorative point of view, water acquires a particular significance as a physical system closely interrelated with the solid and gaseous phases of the soil.

The water present in the soil is studied as a mobile system in which the combination of the constituents, viz, gases and dissolved substances, alongwith biological processes subjected to change with the soil medium, plays vital role.

The atmospheric water is received in the form of precipitation (Rainfall), affects plant's over all development directly, but this water actually is not so important and as vital for plants growth and generic development as the water found in soil-subsoil horizons. This water is generally found in mobile condition in various forms and responsible for the plant's growth and development.

In reality, most of the running water over surface, percolates under neath layers of sub-soil zones and forms a water regime. Thus, in natural process water attains aggregatory to micro-scopic form and exert influence on the generic process of plants.

All plants have certain environmental requirements that must be met, if they are to survive, in due course of time. That is why, specific plants species are found with environmental adaptability in different parts of the world. Some plants (vegetation) are found in very moist climate, known as 'Hygrophytes'. Those adapted to drought are called 'Xerophytes,' and plants with the ability to adjust themselves with the existing conditions are known as 'Tropophytes'. But, above all, plants some way or other, thrive on water. Water is the crucial factor, varying significantly in volume from place to place and in time to time. It provides the medium for the processes of plant growth in the form of moisture and transpiration, and thus, soils play vital role in making this process a complete cycle for the generic development of plant growth. If the water demand, due to evaporation from soil and transpiration from plant surface, is not met, the wilting process may go on in advance and eventual death occurs. Thus, it is the temperature and precipitation which set the conditions for the plant growth more suitable or unfavourable. As we know that drought is a condition where water is needed in excess than the available of moisture. This sets for the alternatives of either abandoning agriculture or to provide water artificially. On the other hand, artificial supply of water is the only answer to agriculture, if is to be carried on. Thus irrigation becomes an important factor.

Irrigation in all its phases, is more emphasised and studied in all the agriculturally based countries, but the different environmental elements like water, air, and temperature regime of the soil, water capacity, impermeability and water budget in soil etc. are less perceived into consideration while watering the crops. All these factors, combine-together form essential conditions for the plant growth and generic process. The time require to achieve maturity is also a function of the length of day or photoperiod. Certain crops receive water in the day, reacts in the process of their growth with the presence of certain amount of heat and sun-light whereas at night, they accumulate Nitrogen (N_2)and other trace elements to complete hydrolysis process alongwith chemical and biological processes. If the watering process in such a critical moment is reversed i.e. watering in the night instead of day, the whole process is adversely affected with crop damage and yield reductions. The effect of water in soil is seasonal, but crucial one. The temperature of soil is adversely affected by the amount of water and thus water regime of soil changes its quality. The Adequate amount of water regulates the soil moisturity on which plant-metabolism process is accomplished and sustain lives. Though, farmers by experiences have learned much from the nature's behaviour and act accordingly. Even then, in the wake of rapid developmental process, we have exploited the nature to an extent of our self gains, and have forbidden natural process.

Of all the characteristic forms of water, movement of water has a greater role. Movements are associated in all the forms i.e. gaseous, solid, and liquid, which enable plants to get water. The amount of water depends upon the phases. Thus, the movement of water determines the quality and types of plants in different regions. The most prominent role of water is found in liquid forms which runs through surface and underground layers. Generally, this phase is known as mobile form. We can not exactly trace-out the under ground running of water, but on the surface it is managed and controlled for crops and plants.

This technique of management is well known as the irrigation; Though, irrigation is a technique of supplying water to the plants or crops, but management of water is a problem in both ways viz; if it is ill managed it causes havoc in the form of flood and drought. Flood occurs mainly due to mismanagement of run-off water over surface and drought occurs due to excess evaporation than the moisture. Thus, both the conditions require greater precession in protecting water from its losses. The better management protects water from its losses. Therefore, irrigation is considered a part of the water management through various measures. This involves the better utilisation of water for plants.

Water is the basic resource for agricultural production. Availability of water and application of right quantity of it at the right time are vital for profitable results from agriculture. In view of the importance of water and its better utilisation for agriculture, development of irrigation has become an issue of inescapable importance in Indian planning. In order to reduce huge water losses occurring in water logging & bogs etc. are problems of paramount importance to check the wastage of water through various means. Irrigation being the key input in agricultural production a large amount of public funds are involved in irrigation projects of all kinds; big medium, and small. In different five year plans the irrigation potential was increased from 22.6 million hectares in 1950-51 to 65.7 million ha. in 1983-84. But the agricultural production could not be raised as per expectations and the gap between irrigation potential created and its actual utilisation widened over time. At the same time the outlay needed per ha. for irrigation potential increased tremendously. Therefore, in the sixth plan period, the emphasis was placed on the better utilisation of the created water potential and the existing potential rather than creating new ones. With this view in mind, Govt. of India established the command Area Development Authorities in major irrigation projects to undertake various on Farm Development activities to improve the efficiency of irrigation water. But the improve the efficiency of irrigation water. But the performance of this scheme undermined because of some reasons like: (i) ambiguity in various concepts used in policy measures and guidelines, (ii) lack of right perception of the policy measures, (iii) non-acceptance of realities, (iv) lack of knowledge about technique and new methods, (v) non-identification of proper research fields and (vi) social constraints.

The policy perspectives and plan envisaged under major and medium irrigation projects have proved to be enormously costly increasing staggering losses and in some cases they are undesirable from the ecological construction point of view. Thus, the concept of irrigation potential and irrigated area etc. seems to be misconceived. So far the minor irrigation schemes are concerned, their role in public sector appears to be more satisfactory as they are less capital incentive and can be constructed within a narrow time frame work, while the potential created is put to virtually immediate use.

Out of the country's cultivable area of 186 million hectares, the ultimate irrigation potential from all major, medium, and minor irrigation schemes is estimated to be 113.5 million ha. of which 58.5 million ha. is from major, and medium schemes and 55 million hectares from minor irrigation schemes.

However, the question is whether we can afford to go about to reach this goal the way it did till now. For the purpose of better land and water management, during the sixth plan, The Command Area Development Programme (CADP), which included field area programme such as field channels, land levelling and waterbandi, could not materialised in time, because of detailed knowledge and information on the extent of water logging in 30.5 million hectares of land where the irrigation potential has been already created by major and medium projects. That there is no precise knowledge about how much and which area need field channels. These are the practical problems facing our country in the plan-period, today.

Besides, irrigation potential created during the plan-periods, the point of irrigation technology and mechanism is more important, but less perceived by our planners.

At present, there is not a problem to provide water to the fields, but the question is whether we are getting the proper returns of input values. In some areas there is a problem of over-flow or flooding of water while some areas are facing drought. Where water is available, we provide excess amount of water to crops which is actually not required. It is a case of overfeeding. While, in same areas we are not providing adequate amount which means ill feeding. This sort of imbalance of watering the crops is more important from the irrigation technology and mechanism point of view.

The present attempt is thus, to focus upon the problem of irrigation technology & mechanism and to see how crops may adequately be provided water of their requirement. Thus, irrigation from traditional method to modern technology and watering through mechanised way are the focal points of the present discussion. In doing so we may prevent water losses through various means, and control it for the better utilisation in order to maximise agricultural production.

With this view, some technical issues related to soil & plant, water management and plant, the conditions of Aridity limit and humidity for plants, irrigation techniques and mechanism etc. have been dealt with to find a solution for harnessing better relationship between plant and environmental elements, in the foregoing units.

TWO

Resource Background

Geography of a region has a profound impact on the general irrigation systems and the lay-out of the water channels, distributaries and sub-channels. It is physiography, climate and soil which determine the direction of lay-out of a drainage system on one side, and the human endeavourance decide the direction of development on the other. In this respect, irrigation system is a result of interaction between natural phenomena and human responses. Thus, both the factors, Natural & Human, play vital role in shaping irrigation system.

Human responses are no more less significant in deciding the general lay-out of water channels and their distribution. Social systems and organisational factors contribute much more towards the watering problems. Land use systems; ownership, tenurial system, and size of holdings are the essential conditions for the better management of agricultural land and its potentiality and thus, provide the future projections of land development activities based on the water management.

Geographical Location

Geographically, Rajasthan state is located in the north-western part of India stretching between 23°3' N to 30°12'N latitudes and 69°-30' E to 78°-17' E Longitudes, covers an area of about 3,42,239 sq.kms.* It forms a rhombus shape with north-south and east-west diagonals of 784 kms. and 850 kms. respectively. Thus, it constitutes second largest state in the Indian Union in area, with a population*1 of about 3,41,08292(Fig. 1).

---

1. *Census of India, 1981, Govt. of India and Basic Statistics, Directorate of Economics and Statistics, Jaipur.

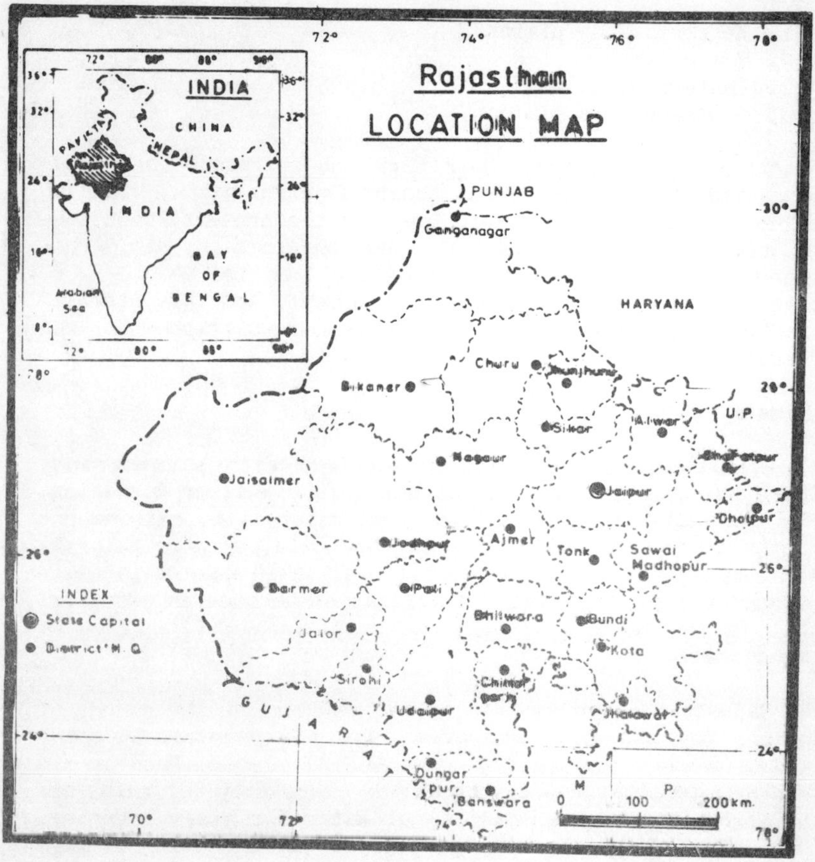

FIG.-1

Physiography

Rajasthan state characteristically is known for its desert in the west and the Aravallis in the eastern flanks of the desert. Thus, physiography is well defined by the geological old structure of pre-cambrian rocks to complex Bundelkhand Gneiss in the south-eastern parts. The Aravallis are well marked in between Desert in the west and Hadoti plateau in the south-east. The usual trend of Aravallis is south-west to north-east, dividing the state into two clear cut physio-graphic Zones. On the basis of the existing relief features this state has been divided into the following physiographic divisions:-

1. Western sandy plains
2. Aravallis
3. Eastern plain and
4. South-eastern plateau.

These divisions further form sub-divisions into Arid, semi-Arid, Luni basin, Banas basin, Chappan basin, vindhyan and Deccan plateau etc. etc. But, for the present study, the importance lies in the fact that how physiography has determined the general conditions of the land and people, more particularly for the economic development. And thus, irrigation comes in forefront of all the resources trapped for the overall development.

Climate

Climatically, the state is endowed with physiography which has provided a distinct but unique weather condition. The seasonal variations of the temperature i.e. extreme hot months in the desert-area to extreme cold, erratic rainfall and humidity are the true conditions, which make this state unique in many respects. But irrigation becomes an essential fact in the background of these variations and vagaries of the climate.

Rajasthan experiences very hot weather in the month of April, May and June and even July. Temperature in these months touches the mercury upto 49°C at places. Even in mid monsoon, temperatures may break the continuity of rains all of sudden and causes scarcity of water for years together. Thus, most of the months go without water causing overall dryness in the soil. The diurnal range of temperature is about 14°C to 17°C in the west of the Aravallis and in other parts of the state also, heating exposes rock surface so intensively that no water is able to preserve in soil . It causes heavy losses of moisture through evaporation and even transpiration.

The distribution of temperature conditions and the rainfall is well marked in Fig-2 which gives the overall picture of the state's climatic conditions.

Vegetation

Vegetation is predominately a result of physio-climatic conditions of a region. In other words, vegetation represents a true colour of terrain and seasonal variations of temperature and precipitation. Whether it is xerophytic or hygrophytic, provide a basis for understanding of a regional development. Rajasthan has a great variety of natural vegetation ranging from scanty xerophytic in the Wn. arid region to mixed deciduous and sub-tropical ever-green forests in the east and SE of Aravalli range. The area covered by these forests is about 2088 thousand ha. constituted about 6 percent of the total area of the state.

TABLE:1

Distribution of Temperature and Rainfall in the Districts of Rajasthan

District	Temperature in 'C'	Average (1981) Rainfall in cm.
Ajmer	23.5	52.73
Alwar	25.0	61.6
Banswara	21.5	92.2
Barmer	26.0	27.7
Bharatpur	26.0	67.1
Bhilwara	24.0	69.9
Bikaner	24.5	26.4
Bundi	- -	76.4
Chittor	23.5	85.2
Churu	23.5	32.5
Dungarpur	25.5	76.1
Sriganganagar	25.0	25.4
Jaipur	25.0	54.8
Jaisalmer	25.00	16.4
Jalore	26.00	42.2
Jhalawar	NA	100.5
Jhunjhunu	23.5	44.5
Jodhpur	25.00	31.8
Kota	27.00	88.5
Nagaur	25.5	38.8
Pali	25.5	49.0
Sawai Madhopur	23.0	68.9
Sikar	24.0	46.6
Sirohi	18.0	63.8
Tonk	25.00	61.4
Udaipur	22.5	62.5
Rajasthan		58.6

Source: Basic Statistics, Directorate of Eco. and Stat.,

Rajasthan, Jaipur, 1982.

The principal types of forests are classified as :-

1. Dry teak forests
2. Anogeissus pondula forests
3. Mixed Deciduous forests
4. Boswellia forests
5. Butea forests
6. Tropical Thorn forests
7. Sub-Tropical Evergreen.

Out of the seven above mentioned types, first three and the last category of forests belong to the Hygrophytes, while the sixth type is a typical xerophyte, whereas, fourth and fifth type falls in Tropophytes which is adaptable to changing conditions (Fig-3).

Human Resources

Human resource plays vital role in shaping economy of a country in a particular direction. To this fact, population, in its manifestations of sex ratio, occupational structure and as a labour (force for industries and more particularly for agriculture) is considered an important element. India is basically an agricultural country where, about 70 percent of the total population is engaged in it. Agriculture as a largest occupation, depends upon the irrigation which is manned by human and animal forces. In this respect, human and animal power, constitute a vital link in the economic development of a region.

Rajasthan state is accounted for about 4 percent of the India's population i.e. 341 lakh, according to the 1981 census, of which 71 lakh or 20.9 percent lives in urban areas and 269.7 lakh or 79.1 percent in rural.

The distribution of rural population by working groups in sex ratio is given in the table 2.

Those workers who are engaged in agricultural and other allied activities are partly engaged either in lifting water or supplying it through channels and thus, form an important part of the irrigation system. The other aspect of water regulatory system is associated with the animal power as main source of draught, drawing plough, bullock carts, and even charas and Rehant (Persian wheal) to pull water from the wells. Thus, animal force also constitutes an integral part of the whole system as a mechanism. Rajasthan holds about 413.5 Lakh livestock population according to 1977 census. If all the working categories of human and draught force is converted into equivalent to horse power(H.P) units on the basis of the International equivalent power unit table deviced by the I.G.U., the following picture emerges in the state.

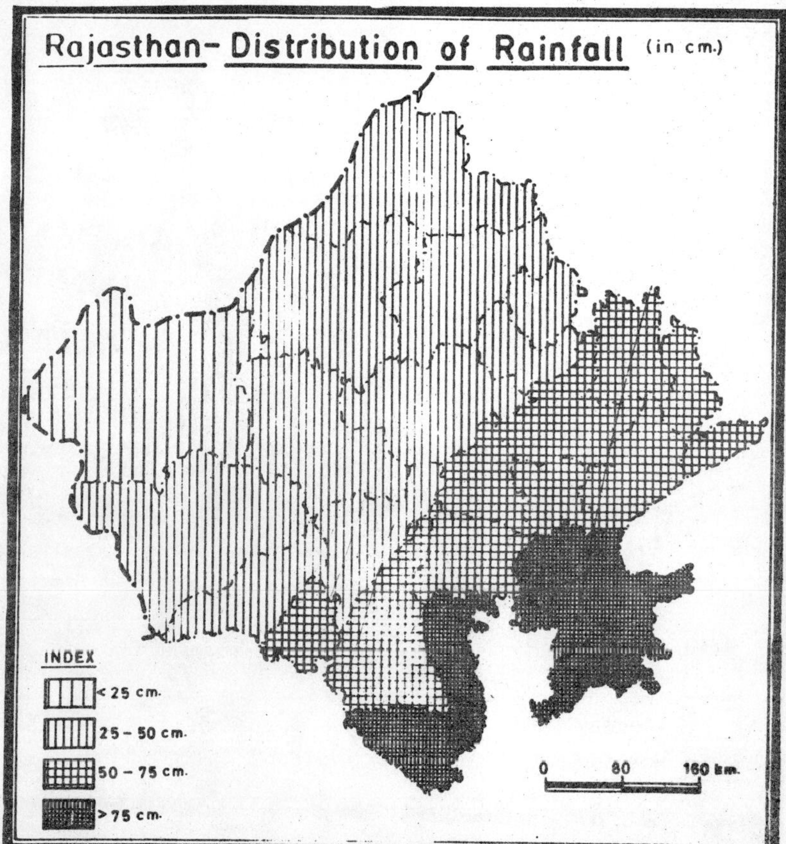

FIG 2

All these power units deployed in agricultural operation are important from the irrigation point of view as main source of irrigation along with driven machines like pumpsets, pumping of water from the wells, tanks, and as pressure builders. Though, the state is reckoned with the traditional source of irrigating the fields in which human and animal labour is deployed extensively, to pull kehant and charas as the most common means of water lifting in the villages even to day. Therefore irrigation practices and methods are well associated with the traditional ways in the vicinity of developed techniques and methods.

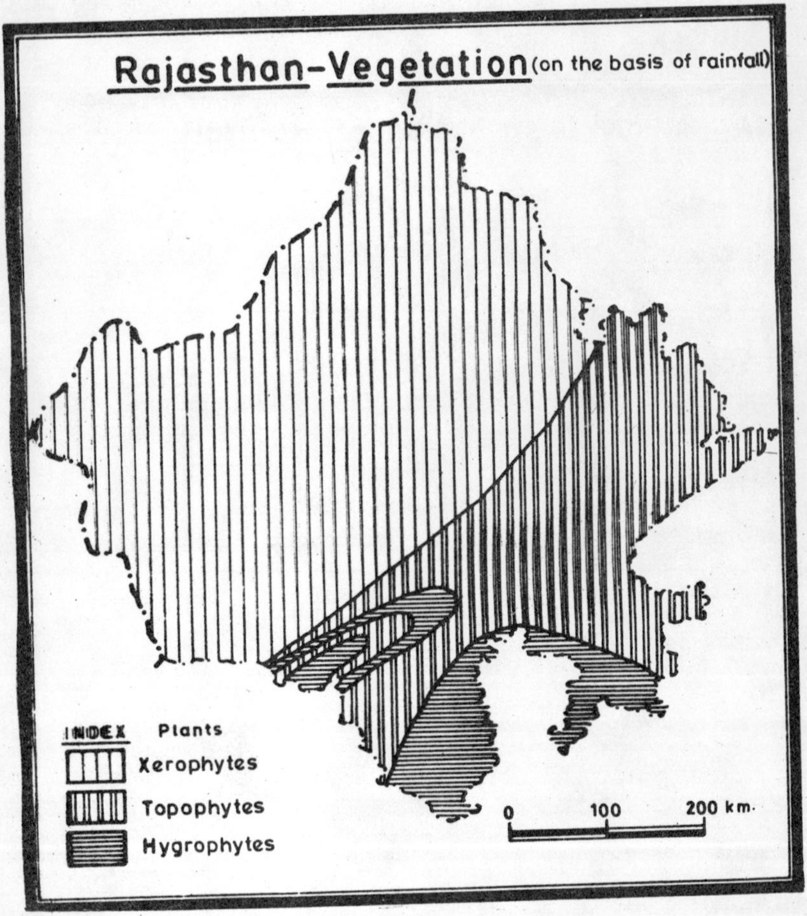

fig. 3

Besides, animal as major source of power, they thrive in accordance with the water availability. Their water requirements also vary from stock to stock specially in milching animal. This aspect also makes a point to be of worth consideration.

Human power in H.P. equivalent - 12.4 lakh units.
Animal power in H.P. equivalent - 3.6 Million units.

Mechanical H.P.Units - 4.4 Lakh units

TABLE-2

Distribution of Working Rural Population by Sex

No. in "000(1981)

Classification	Male	Female	Total	%age to total population
1. Main workers	7,083	1,383	8,476	24.9
i Cultivators	5,239	1,003	6,242	18.2
ii Agricultural labour	479	231	710	2.0
iii Housing Indus. servicing & repairs	216	31	247	0.7
iv Other workers	1,149	128	1,277	3.8
2. Urban main workers	1,722	153	1,875	5.5

Source: Directorate of Eco. & Stat., Raj., Jaipur. Basic Statistics 1982.

GENERAL LAND USE IN THE STATE

Land use patterns provide the basis for development. From the irrigation point of view, agricultural land use categories pave the way for intensiveness. The intensity and diversification of land use changes affect the spatial dimensions of agricultural infrastructure and more particularly the extent of irrigation.

The general land use classification in the state is found as under:-

TABLE-3

Land Use Categories

Categories	Area in ,000 Ha.	1981 % to Total Area
1. State Geographical area	34,226	100.00
2. Forests	2,087	6.10
3. Land not available for agriculture		
(a) Land put to nonagricultural uses	1,507	4.40

(b) Barren and uncultivated land	2,917	8.52
4. Land not available for cultivation(excluding fallow)		
(a) Permanent and grazing land	1,833	5.36
(b) Land under misc. tree, crops, &	24	0.07
(c) Culturable waste	6,415	18.75
5. Fallow land		12.1
(a) Current fallow	2,085	6.0
(b) Long fallow	2,089	6.1
6. Net sown area	15,267	44.6
7. Land sown more than once	2,082	6.08
8. Total cropped area	17,349.6	50.71

State Abstract, Directorate Eco. & Stat., Jaipur, pp.39-48, 1982.

From the above table it may be brought out that the state is very poorly located in agricultural land uses. It has about 285.5 lakh ha. as cultivable land or 83.3% of the total land, but actually used under cultivation is about 173.5 lakh which is 60.7% of the cultivated land. It means near about 40% of the arable land is not being used for cultivation because, of the irrigation facilities. In addition to it, if fallow land and other miscellaneous land is taken into account which comes about 36.1% of the arable land, is sufficient to feed the population. But, in want of water, the land is lying unused or partly used when water is available through rains. The irrigated land accounted for only about 17.13 percent of the agricultural land by all sources, which is insufficient for the state like Raj. The state has about 50.7 percent gross cropped area against the 44.6% of net sown area. It means, most of the net sown area has to depend upon either rain water or has to go along with dry cultivation. Thus, dry farming too, has large areas under it, while extensive area come under rain-fed agriculture in Kharif season. It is clear from the evidence of the area sown more than once, which constitute only 6 percent, that most of the net sown area which accounted for about 44% is not used in succession, due to lack of water or irrigation. It reflects that the remaining 38 percent area of the net sown area being put partly in rainfed agriculture and partly in dry farming. Thus, if irrigation is made available to the farmers the entire area alongwith other arable land may be put to intensive cultivation.

Size of Land Holdings

The size of land holding from the irrigation point of view, is more significant than merely a social system. The land pieces and fragmentation play vital role in irrigation technology and mechanism. If the land is fragmented and found in pieces, the incidences of water losses are greater than a large chunk of field, where water is better utilised with minimising the losses. Thus, consolidation of land holding programme was a step towards preventing water losses, but the scheme was jeopardised due to social constraints.

In fact, "watering of crop is a package deal not a partial agreement. But in traditional country like India, it is a partial deal rather than a package agreement." This holds the view that every individual farmer in India has his own arrangement of irrigation system for piece of his holding, irrespective of the size. Though, it is indisputable that on an average, majority of the farming class holds the small or marginal holdings of 1 to 2 ha. counted nearly 70% of the total operational holdings depend on partial irrigation. Therefore, irrigation potentials created for them are meeting with failure at all levels. Thus, irrigation techniques and mechanism may find a way out for the crucial problem of such wastages.

So far the operational size of the land holding is concerned, it has a very unusual distribution in different size groups. The recent agricultural census data levels that the marginal holdings of below one ha. accounts about 51 percent of the 70 million operational holdings and occupies only 9 percent of the total cultivated land. The small holdings including marginal holdings also, having the size of holding upto 2 ha., account for about 70 percent of the operational holdings but the area operated by this group is only 21%. The third category of semi-medium holdings of 2 to 4 ha. constitute 15% of the operated are under it. While medium holdings of 4 to 10 ha. account for about 11% of the total holdings with 30% of the total cultivated land. The large holdings of more than 10 ha., constitute hardly 4% of the total holdings cover nearly 30.5% of the total cultivated land.

This anomaly of the size groups certainly have the impact on the irrigation systems and methods. As an observation irrigated and unirrigated land with sizegroup, is stated below in the table which provides a clean picture of the state for the future projections and the planning perspectives.

TABLE

Holdings and Area Irrigated by Size Class

Size class ha.	NSS Categories	Wholly No.	Irrigated Area	%age of 3	%age of 4
1	2	3	4	5	6
Below -0.5	Small	119.1	27.5	70.5	22.1
0-5 -1.0	&				
1-0 -2.0	Marginal	74.4	46.9		
2.0 -3.0	Medium	71.0	83.4	13.9	8.5
3.0 -4.0		31.3	6.0		
4.0 -5.0	Semi-medium	20.6	55.8	12.0	31.0
5.0 -10.0		10.9	37.8		
		34.1	175.8		
10.0 -20.0		10.5	105.1	3.5	38.3
20.0 -30.0	Big	1.2	18.5		
40.0 -50.0		0.3	6.9		
		0.1	2.8		
		0.4	66.6		
	TOTAL	374.2	687.8	100	100

From the above table recurrences may be drawn that the percentage of number of holdings and the area irrigated by size of holdings, are quite in contrast to their categories. So far the number of holdings are concerned, 70.5% belongs to the small & marginal size of their total holdings whereas big holdings of more than 10 ha. are found with only 3.5% of their total holdings. While the area irrigated in their size groups, is accounted for 38.3% in the big size holdings and small holdings counted about 22.2% only.

The anomaly of distribution of irrigated area according to the size of holdings may be expressed in the words that the larger group of holdings of less than 2 ha. of land, accounted for about 70.5 percent of the total holdings but only 20.2 percent of the total area is irrigated. Whereas, the smallest group of big holdings of over intire holdings and get irrigation to 38.3 percent of the total area. The remaining size groups of medium & semi-medium categories also get favourable response in irrigation. Though, the semi-medium category of 4 to 10 ha. rank next to big size in irrigation by 31 percent. But, Medium size too, is a sufferer.

Therefore, it proves the hypothesis that the small farmers are partially benefited while the big farmers are getting all the benefits of their big chunks, so far the irrigation is concerned.

Jaisalmer district in Rajasthan represents an interesting case of the biggest size of land holdings i.e. 103. ha., in the Indian union. The small size of land holdings is more concentrated in the eastern Rajasthan. While wn. parts are represented with big size. It is in view of the irrigation, that the size of holdings play vital role in deciding the use of the land i.e. whether it is within the range of irrigation or not. Out-field and in-field systems are accorded on the basis of the range of irrigation water alongwith fertilizers' use.

THREE

Soil Responses to Plants

Soils are the major sources of material wealth on which mankind has depended for centuries to obtain food. Soil according to the Russian Scientist, is a body subjected to a natural and historical development, which came into being on the surface of the earth as a result of a complex combination of the interactions of rocks, the organic world, the climate and the production activities of man. In this context, the formation of soil, composition and structure are the essential conditions of the soil components which play vital role in the production activity of plants and crops. All these forms are a complex phenomena in natural process under which soil passes and acquire a particular shape and characteristics. Soils acquire physical, chemical and organic properties while under going transformation of natural processes where water plays an important role in such transformation.

Water is the most-important condition for the development and fertility of soils. Water acquires a particular significance as a physical system closely interrelated with the solid gaseous phases of the soil and plants. Thus, it is the form of water in the soil, which immediately reacts and behaves in a particular manner responsible for plant's growth and development, governs the direction followed by soil formation and fertility. Thus the regulation of water very much depends upon the water regime, and water balance of soil which constitutes a highly significant task in confronting agricultural development.

Simply watering the crops does not solve the problem, but it is the soil through which water is required to plants. What amount of water, in what forms is being received by the plants, whereupon plants growth depends, is the quest. Therefore, one has to take care of soil and water conditions before going to supply water to the field.

It becomes more crucial factor when the supplied or released amount of water is not reached to the plants and a substaintial part of it is lost in atmospheric circulation through evaporation or transpiration or even percolation. Considering water as a scarce element, obsolete and faulty irrigation methods are to be considered important in time to come, alongwith the ground conditions.

There is a complex interrelationships between soil and water. It is found in various aggregatory conditions and in different forms in soil. Plants suck water from the soil's aggregatory colloids and not directly as supplied through stream (channels). Soil provides a good medium for the generic process of plants.

Forms of Water

Water in the soil is found in the following forms:-

1. Chemically combined water of
    (a) constitution
    (b) crystallisation

2. Physically bond water
    (a) tightly bond (hygroscopic)
    (b) loosely bond (film water Molecular)

3. Capillary water
    (a) abutment water (pendular)
    (b) funicular
    (c) perched
    (d) supported
    (e) Mobile

4. Gravitational water
    (a) leakage
    (b) supported
    (c) percolating
    (d) ground water

5. Water vapour

6. Water in the solid state.

The various forms of water in the soil are controlled by

the soil types. The characteristics attributed to soil types are assigned by formation, structure and texture of the soil which absorb and contain water in various forms. Thus the capacity of absorbing water and retaining it for an indefinite period, depends upon the soil characteristics. The water controlling capacity in different soil groups may be seen in the table below:-

TABLE 4

Hygroscopic Water Content in Percentage

Mechanical composition of soil	Hygroscopic water as % of weight of absolutely drysoil	Maximum Hygroscopic water in % age
Sand	0.5 - 1.5	1.5 & more
Light loam	1.5 - 3.0	3.0 - 5.0
Medium loam	2.5 - 4.0	5.0 - 6.0
Clay	6.0 - 8.0	8.0 - 10.0
Peat	18.0 - 22.0	-

Plyusnin, 1.1 Reclamative soil science, Mascow, p.124.

Fluctuation of maximum hygroscopic moisture depends on the amount and composition of absorbed bases and soluble salts. For example, one salined clayey soil is capable of holding 27% of maximum hygroscopic moisture and after leaching only 14.4% while saline loamy soil before leaching held 16.5% and after leaching only 6.7%.

Thus, the water holding capacity of soil is the capacity to contain and retain the maximum amount of moisture upon the given conditions. It depends upon the forces which exert their effect in soil and on the factors of soil formation. It can also be seen that the range of available active water lies between moisture of wilting and maximum water holding capacity of the soil. In poorly drained soils or areas of inadequate irrigation, when the level of ground water is high, soil moisture may come to exceed limit field water capacity, with attending extremely poor aeration. In such conditions plants reacts at first by slower growth which is followed by death. It is of the great significance that at what depth roots are distributed. A shallow penetration of the root system on humid poorly irrigated soils may prove disadvantageous.

In places where, in the first half of the vegetative

period, the top layer of the soil dries out, plants in the beginning, develop a shallow root system and adapt themselves to very moist conditions suffer from drought in the second half of the vegetative period, simply because roots are unable to follow the slowering of the level of capillary fringe. This applies to sprinkling irrigation with frequent light watering at the beginning of the vegetative period. Subsequently plants suffer from shortage of water , whose roots remain within capillary fringe, cease to be affected by change of the moisture content in the top soil horizon.

With the assumption of the lay down principles of the soil and water' from the irrigation point of view, the soils of this state may be taken into consideration on the basis of their occurrence, chief characteristics and suitability for cultivation. Ray-Chaudhri (1984) has divided soils into seven major groups.[1]

1. Desert soil
2. Grey and baron (desert) soil
3. Red and yellow soil
4. Ferruginous red soil
5. Mixed red and Black soil
6. Medium black soil
7. Alluvial soils.

Desert soils occupy larger area (3/5) in the state. Right from the Wn. Aravallis to the border areas with Pakistan, is covered by sand, hillocks and outcrops, constitute the aeolian sand, with debris commist of caO to $CacO_3$ known as kankar bed formation. The general characteristics of the desert-soils are immature structureless, very coarse in texture, low in water holding capacity and low in base exchange property. Poor moisture retentive coupled with lack of binding material and very meagre bacterial activity make the desert soil highly unproductive. This has high proportion of mineral plant-nutrients, especially calcium, magnesium and potassium, but the contents of organic matter and total nitrogen is very low. These facts lead to the inference that deficiency of moisture rather than of plant food elements is the main cause limiting the fertility.

---

1. Ray-chaudhri; S.P.Land Resources of India, Indian soils and their classifications, Manager of Pub., New Delhi, Vol. 1964

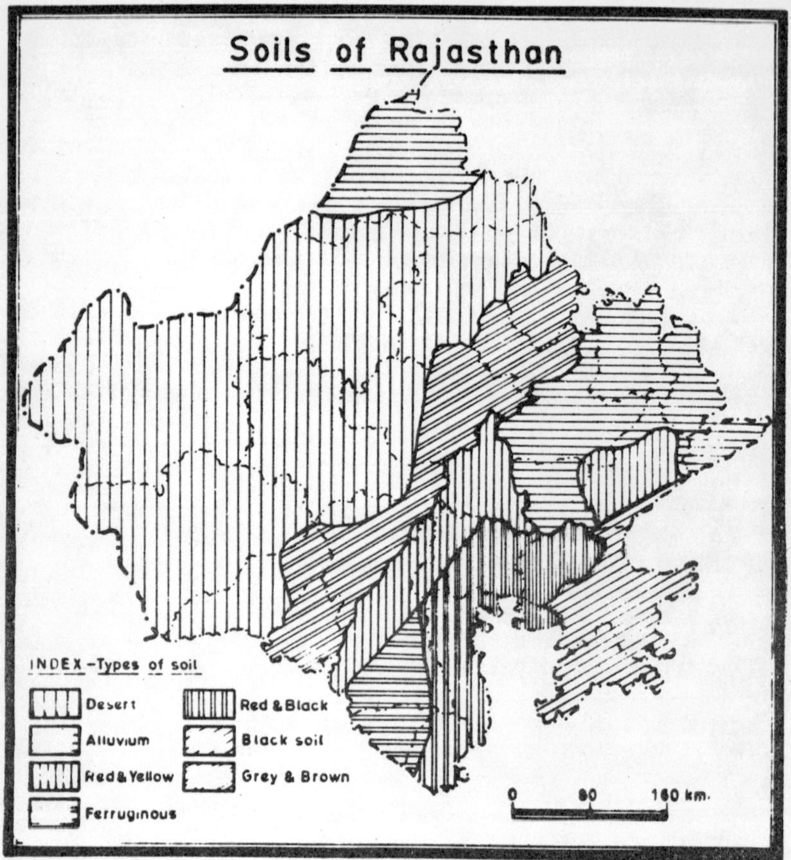

Fig. 4

Same is the case with grey and brown desert soils occur in the peripheral districts of the desert in the eastern margins. They are composed of fine soil sandy in nature with a poor moisture holding capacity and good in permeability. These soils have good contents of phosphates and nitrates which enhance fertility on getting water.

Red and yellow soils are the composition of silty loam and silty clay, which is poor in carbonates and humas but have a good moisture holding capacity. Thus Udaipur, Bhilwara and Ajmer Districts, are enriched with a bit longer moisture period. Ferruginous red soils are having lighter texture, porous, and friable structure, with lime, potash, iron oxide and phosphorous contents. All this makes the soil semi-fertile on getting water.

Most of the Southern districts of the state (Udaipur, Chittor, Dungarpur, Banswara and Bhilwara) are located in Red and Black soil region which is an admixture of many mineralised contents like iron oxide, magnasium, potash, calcium, nitrogen etc. The P^H value of the soil is natural to alkaline. The Black soil of the plateau region in the SEn. part of the State constitutes varying degree of texture ranging from clayey to loamy. It has the high capacity of holding the water,even at the times of high temperature.

Characteristically, alluvial soil is considered the best and most fertile soil. It is because of the varying degree of textures which makes it clayey, loamy and sandy. The admixture of these textures makes the soil most permeable, with humas contents. But found with deficient in phosphates, lime and variable contents of nitrogen. This soil is spreaded over the north and NEn. parts of the state mainly in the districts of Alwar, Bharatpur, Jaipur, Tonk, Sawai Madhopur and Sriganganagar (Fig. 4).

The study of all these soil groups provides a fundamental basis for the better understanding of the water requirements and management which helps in general layout of the channels and distributaries as a part of irrigation system.

FOUR

Cropping Pattern in Relation to Water

Plant & Water Requirement

Agriculture is essentially a combination of processes designated to promote, artificially, a favourable environment for growth of plants. Its practices are so diverse and varied in nature as to include drainage and irrigation to control the limitation of the moisture, artificial sheltering to control temperature, and sheding to control sun light. All crops have climatic limits of economic production which can be extended by plant breeding and selection as well as by cultivation methods. In exerting influence on crops, climatic factors are closely interrelated. Temperature and rainfall are the factors, directly affecting plant growth and determining the efficiency of crops. But, the limitations of water in the form of moisture are always there, though, the variation of temperature and rainfall exists.

Inspite of all that each plant family (species) has its own water requirements as it differs in growth cycling of a plant. In a complete vegetative period of a plant, various forces contain to accomplish photosynthesis and metabolism process of plant genesis. These processes depend on the soil characteristics and the amount of water. Mineralised soils react in accordance with water and prepare the conditions of food for plants. Some mineral absorbs more water while some less.

Thus, water requirement varies from plant to plant, and crop to crop. In some cases, the water requirement in a cropping period is stated below:-

Wheat	3 to 6 times in a cropping period
Barely	2 to 4 times " "

Maize	2 times " "
Jowar	2 - 3 times[if no rain in Kharif
Bajara	2 - 4 times[" "
Rice	Maximum at all times except harvesting period
Sugarcane	10 times
Cotton	5 - 9 times
Lucerne	Weakly in summer 15 days in winter

The water is utilised differently by the plant during the different growth phases and in a vegetative period. The yield experiments with rice and wheat as a function of the time at which a physiological water deficit occurs, show the following results. In first case experiment was carried on rice crop in Philippines which revealed that:-

adequate water supply 6.2 t/ha 8.1 Kg/mm water
no water from 41 st. - 81st lay 44 t/ha 5.6 Kg/mm water
no water from 63rd to 102nd day 2 t/ha 2.5 Kg/mm water
no water from 63rd day to,

harvest -- 0.5 t/ha 1.5 Kg/mm water

In the second case, experiment was carried on wheat crop by the department of central Aridzone institute (CAZRI). Jodhpur (Raj.), and obtained the following results :

TABLE-5

-Line	Yield	Water requires under management levels	
		Sub optimal	Optimal
No.	Kg/hs	----------cm-------------	
1.	2,272	29	18
2.	3,150	46	30
3.	3,511	56	35
4.	3,909	82	42
5.	5,260	imaginary	79
6.	5,460	imaginary	98
7.	4,122	82	--
8.	4,529	--	56

Source: Optimisation of water use of crop production, p. 25, Jodhpur (CAZRI).

It is well known fact that Rice crop require maximum amount of water, right from the sowing to the ripening period, as it is a crop of wet climate. Then comes sugarcane next to it, which also needs maximum amount of water. It is noted that at the beginning of the seedling time fields should be filled with water about 12 to 15cm., but it needs irrigation at the interval of about 15 to 20 days at the time of its growth. Otherwise the growth may banish to decline the quality and hence, the production. Thus, it is a crop of damp climate with sufficient temperature.

In some of the fodder crops lucern is the most flourishing plant in the field. But it depends primarily upon the water at an interval of five to six days i.e., weekly. It is the most water consuming plant, having longest root system beneath the earth at the depth of about 1.5 M. Some way or other, all the rabi crops needed water at a set rotation, depending upon the crops which may reduce the extent of irrigation period, e.g., wheat needed irrigation at the length of about a month's interval, (though, all these watering periods are more or less depend upon the existing temperature) whereas, lucerne, rice, or sugarcane reduces to a week's interval.

In kharif season i.e., the crops harvested in summer just after the rains, needed very meager amount of water as they get sufficient moisture from the soil regime. But in areas where evaporation rate is higher, moisture is not retained in soil for a longer period, and thereby, crops are required to supply water through irrigation. Though, there are crops which consume less amount of water, means their water requirement is very limited in their generic process. These crops are well known as maize, Bajra.

Thus, in areas of long days of dryness, or inadequate moisture in soil, Rain-fed agriculture and Dry-farming are practiced. These crops are well associated with seasons as per their water requirement, and thus farmers have set harvesting periods in such areas.

Cropping Pattern

Agriculture essentially is a way or practice of raising crops on soils. Soil combine with water is a factor of production limiting to soil types and amount of water. Thus, crops vary from place to place and in time to time. Actually it is the generic process of plants accomplished in accordance with the set temperature and adequate amount of water. That is why two cropping sets of seasons i.e. Kharif and Rabi, are practiced during summer and winter respectively. Crops which are set to thrive in their generic process are never cultivated against their set seasons, and if natural process is not followed they are deemed to yield anything. Thus a set of cropping seasons is found in all over the world. They are:-

1. Crops grown in the Rabi season: Wheat, Barley, Sugarcane, Gram, Mustard Soyabean, Beans etc. etc.
2. Crops grown in the Kharif season: Maize, Jowar, Bajra, Pulses, Cotton, Till, Groundnut, Mung, Methi Urd, etc. etc.

The irrigated area under Rabi and Kharif seasons in respective crops, is shown in the table below:-

TABLE-6

Area Under Rabi Crops

(in '000 ha.)

Crops	Total	Irrigated	%age to Total	Area not irrigated in %age to total
Wheat	1632	1389	85.1	14.9
Barley	409	302	73.8	26.2
Rice	170	73	42.9	57.1
Sugarcane	29	28	96.5	3.5
Gram	1227	391	31.8	68.2
Other Rabi Pulses	28	12	42.8	57.2

Kharif Crops

Crops	Total	Irrigated	Rainfed area in %age
Bajra	5035	145	97.1
Jowar	1006	10	99.1
Maize	899	209	76.7
Small Millets	47	2	95.7
Cotton	357	317	11.3
Other Kharif	1758	6	99.7
Groundnut	212	16	92.5

It is well known that Rabi crops needed irrigation whereas Kharif does not. The above figures for the Rabi crops reveal that about 2/3 to 1/3 area of the respective crops remained un-irrigated when water is needed. This means the remaining area under respective crops have to leave either as fallow or for dry farming, as it does not get irrigation.

Same is the case with Kharif crops, though it requires no irrigation, but some perennial crops and vegetables needed irrigation especially in times of long intervals of precipitation. Thus, more than ninety percent of the area falling in Kharif crops, is termed as rainfed. The cultivation in such areas in known as 'Rainfed' agriculture.

As regards the cropping pattern, Rajasthan can distinctively be marked in two major crop-regions viz; West-coarse grain region, and East-multigrain region. Wn. Rajasthan (West of the Aravallis) is primarily dominated by Bajra, as a single mono-crop in majority of the districts followed by Jowar and Pulses in some parts (Fig.5). Sriganganagar district provides a case of exception, where multicropping pattern is witnessed. The eastern Rajasthan is dominated by a variety of crops in both the seasons, the most common amongst them are Wheat, Barley, Pulses, Cotton, Mustard, Sugarcane, Maize etc.etc..

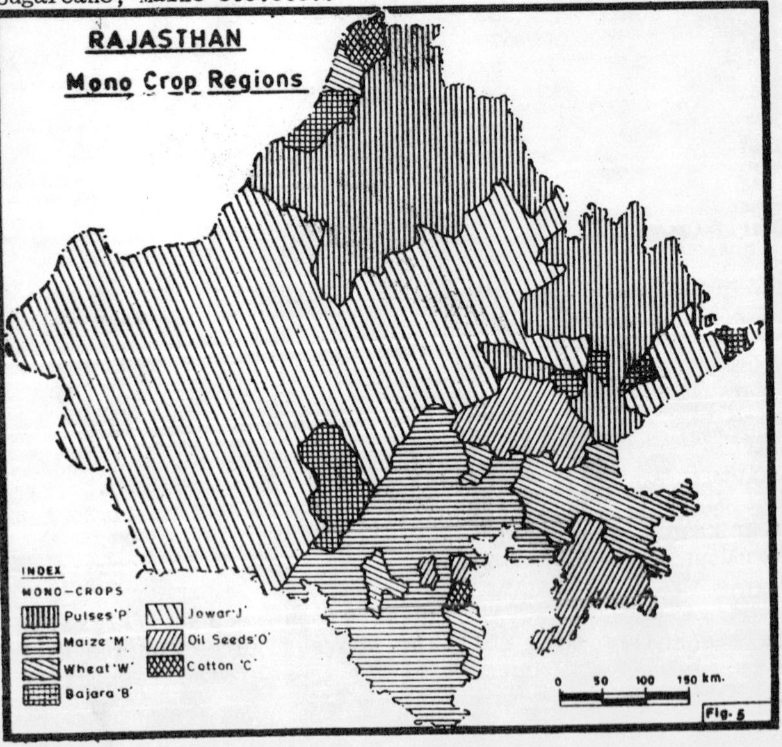

The cropping pattern may best be visualised from the crop-combination regions, so far their importance as well as dietary habits, are concerned (Fig. 6). Mono-crop region is predominantly seen in the Western most-districts of Jaisalmer and Barmer. While, other districts of the desert margins, Jodhpur, Bikaner, Churu, Jhunjhunu, Sikar, Nagaur, and Ajmer are found with two-crop combinations i.e., Bajra and pulses. Three crop-combinations are also found in vicinity of these districts, whereas four crops combinations scatterly distributed in the parts of Alwar, S.Madhopur, Jaipur, Jalore, Pali and Kota districts, Five and six crop-combination comprising of wheat, oilseeds, pulses, bajra, and jowar are confined to the areas of eastern Rajasthan particularly in Chittor, Bharatpur, Sirohi, Udaipur, Kota, Tonk, Bundi and Ajmer districts etc. The most common but commercial crop like mustard is being cultivated extensively in the eastern districts of the state while southern parts specialising in maize cultivation. Thus, the range of combinations go upto the seven in which variety of crops are grown on small patches of plots, which found easy to be irrigated in limited water resources (Fig.6).

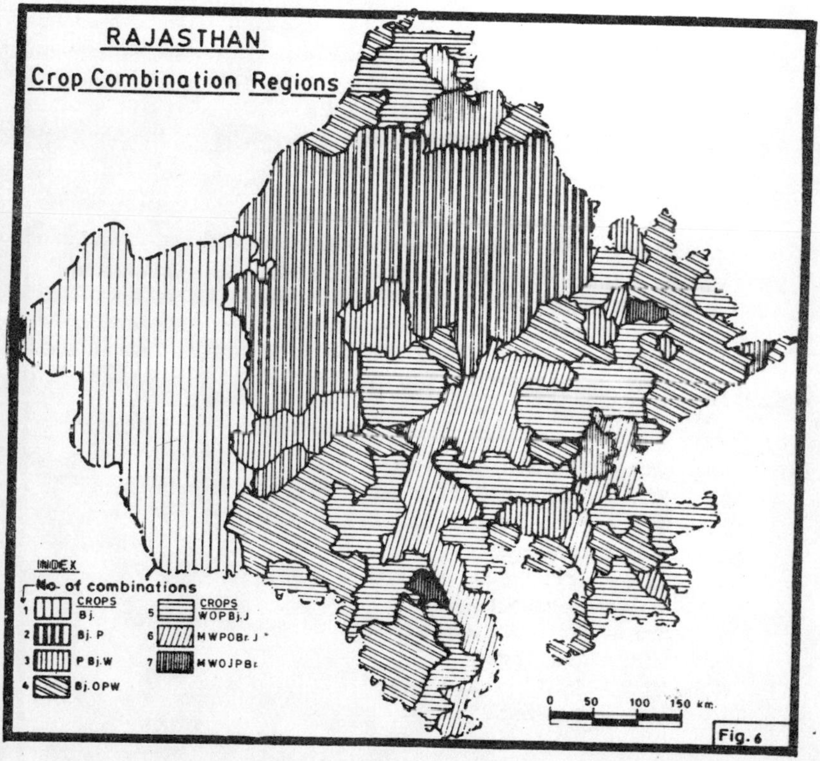

Fig.6

All these combinations are reflective of the regional characteristics well confined to the availability of water. The extent and the intensiveness of cropping is directly associated with the amount of water. Because, we have seen that certain crops require more water particularly in rabi season. Whereas, the range of Kharif crops and the variety is a function of rainfall, directly related to the periods and the distribution of it. Some times the range of harvesting extended alongwith the widespread showering of rains, while it is folded on limiting of the showering. Above all, water is the principal factor in determining crop combinations and their associations of important crops with variety.

On the basis of the extent of irrigation alongwith other factors the state has been identified with six agricultural types (typology) by the author himself, which further provide the insight-into the problems of watering the crops. These typological regions help in understanding the spatial characteristics as well as the regional taxonomy. Because, it is not necessary that spatial dimension is only factor determining agricultural patterns and the production characteristics but at times, social and organisational activities, abide by with traditional ways, social behaviour, customs, traditions, and institutions etc., influence the overall characteristics in turn. Irrigation practices, a way and method of giving water to the crops, are by no means less significant and unaffected by these social practices.

The most interesting fact of these typological regions is that they are marked with some irrigated and semi-irrigated types of agriculture. It means, the irrigation also plays an important role in determining agricultural types. These types ranging from nomadism to the small scale mixed commercial agriculture are located right from Wn. most districts of Jaisalmer to the eastern most Sriganganagar. They are listed as below:-

1. Enn - Nomadic Herding
2. Eth - Current fallow with migration herding
3. Tir - Semi-irrigated, low productive, Semi-subsistence.
4. Tiu - Irrigated, Medium productive semi-subsistence to semi-commercial.
5. Tmh - Semi-irrigated, Medium productive with live-stock grazing.
6. Mmc - Small scale mixed agriculture with crop growing prevalent.[1]

The above types reveal the fact that type 3,4, and 5 exclusively based on irrigation, whether it is fully irrigated or partly that mean semi. The production capacity and subsistence or commercial enterprise system, depends on the water. These types are covered under the districts mostly falling in the east of Aravallis to the SEn. part of the state. In a semi-circular form Ajmer, Sirohi, Pali, Tonk, S.Madhopur, and Kota are characterised as Tir type. Whereas in the south of the upper crescent of Tir, Tiu type of agriculture in Udaipur, Chittor, Banswara, Bundi and Jhalawar form an important part. Sriganganagar district is an exception to all that which falls in Mmc type. All it is because of irrigation. The 5th-Tmh type covers the districts of Jaipur, Alwar and Bharatpur, which again covers the districts of Jaipur, Alwar and Bharatpur, which again depicts the fact that water is a crucial factor in all these activities.

Besides all that the future projection of these districts lies in the fact that they needed more extensive irrigation in future to come, if agriculture is to be thrived well.

---

1. Sharma,B.L. Agrl. Typology of Rajasthan, Pankaj prakashan, pp 124-130.

FIVE

Water Properties and Plants

Perspective of Water and its Uses

When viewed from space the world is really seen a globe of water. Oceans cover 71% of the earth's surface and polar ice and mountain glaciers cover an addition of 3.18% . The volume of water in the world would be sufficient to enclose the entire globe in a layer of 2,718 M. deep. Of all the water in the world i.e., 1,338 billion Km^3 or 96.5% is concentrated in the oceans. The continental water is divided more or less between surface water and ground water. If all the continental water i.e., 47,971000 Km^3 or 3.5% of the entire volume, is evenly spreaded on the surface it would produce a layer of 322 M. thick over all the continients. It is perhaps with this assumption that water is abundantly available on the surface, so we make use of the water recklessly and without any reservations.(Keller, R. Applied Gevg & Dev, Vol.24, 1984)

The figures for the volume of ground water are very unreliable, until around 1970s it was estimated (df. Keller, 1966) that 4 billion Km^3 of ground water was located at the depth of 800 M. in a zone where water can generally be used. But there is a problem to obtain fresh water from such a depth which varies from place to place and zone to zone. Moreover the large resources of fresh water in the under ground can be used only within certain limits. It is estimated that the volume of fresh water in the underground is about 10.5 million Km^3 accumulated from the area of (Catchment) 134.8 Mi. Km^3 at the depth of 78 M. This amount constitutes 0.76% of the hole water resources of the world, and 30.1% of the fresh water resources of the world's estimates. The actual amount of fresh water in respect of total volume of world water is about 35 Mi Km^3 to 1,386 Mi Km^3 respectively.

In contrast to the world resources of water, India is said to be placed in a better position, but so far the fresh water and actually trapped water for irrigation is concerned, it is meagrely placed. The water resources by river systems alone contribute about 55 billion cubic metres, of which surface water accounted for about 45 bi M^3 and ground water is 80 bi M^3; The annual surface that is accounted for about 1897 bi M^3 (Johnson B.L.C. 1980, India, p. 69). But country's present irrigation potential is estimated around 57 Mi hectares, which is to be raised upto 113.5 Mi hectares in the seventh plan.

Rajasthan is poorly presented in Indian Union so far its irrigation potentials are concerned. The State irrigation Dept. has revealed the facts that there are about 709 T Mcf of surface water potential from all the river systems is available to it. But, for the effective use in irrigation is estimated about 500.3 T (Mcf) cubic feet. Thus, Rajasthan has wide reservoirs of water from river systems and even estimated that a large amount of water is available from the huge ground water reservoirs which are unutilised extensively so far. (The natural Resources of Rajasthan, Vol.2, pp.641-43).

Irrigation is a art of artificial supplying of water to the crops. This involves means through which water is supplied on one hand, and the methods by which irrigation is practiced on the other. But, while doing so much of the water is either wasted or lost through evaporation and transpiration. The rate of such losses is measured through total precipitation and potential evapo-transpiration. But the actual losses are much more than what is measured. These losses are accounted for by mis-use of water by human being, through the mal-practices and mis-management. Therefore, how to use water or manage and to what extent are the questions of paramount importance before all of us. Are we going to use water in the same way as we have been doing? Are we getting the water to drink for all mankind in future? Will it be available in the same quality? These questions being asked everywhere with the suspective eyes of the millions of people with thirsty throats, on the earth, to day.

The endangers of the rapid strides of the population growth, all over the world, more particularly in Asian countries, are well realised by the scientists and grim prospects of inadequacy of water for plants and people are perfectly calculated in the light of future prospects, by

2000 AD. But the lowering water table in all parts of the world, increasing demand of water for plants due to excessive aridity limits, and growing pollution in water are the glaring problems of today, and giving timely alarm to both users and applicators of water. Thus, water needs to deal with scientific utmost care and sensible manner.

Water is available in many forms. All the forms are not suitable for plants and people. That has to make suitable to drink and even supply to plants. Some of the forms (quality) are discussed herein:-

Quality of Water

The quality of water depends upon the properties found in water. These properties make the water a different one viz; some times it is heavy or light, some times salty or sweet in drinking. But, it altogether differs in characteristics when tested in laboratory. Thus, the following parameters are fixed to measure the quality of water: (i) on the basis of physical properties and (ii) chemical composition.

The physical properties of water are well ascribed in the form of molecular hygroscopic water, expressed in texture and composition of molecules of water. These characteristics are normally based on the water forms - Film or capillary or gravitational etc. etc. and do not affect crops to a great extent. But, the Electric conductivity (EC), which is the best expression in scientific term of the physical bound water makes a matter much for crops or plants. EC is a principal parameter to determine quality of water for irrigation though, EC is a rate of mobility of molecule or particles in water on meeting with certain chemicals. It reacts on the combination of bases and some trace elements in water. For fixation, certain adjustments are made between cations and anions, on which a definite molecular movement is set e.g., some important cations and anions are given below:-

Ca^{+-}-Calcium, Na^{+} - Sodium, Mg^{+} - Magnasium,
Si -Silicon, and B-Boron^{+} etc.

Anions are :-

Cl^{-}-Chloride, S^{-}-Sulphates, O_2-Oxygen.
N_2^{-}-Nitrogen, F^{-}-Florine, Br^{-} -Bromide etc.etc.

Cations are having positive electrons where anions negative charge (Minus) and thus, fixation of elements is accorded by giving-taking electrons of positive-negative elements. Viz; Na^+ Cl - Sodium Chloride samples from Jodhpur district having Ec value 1.8 to 18.0 Mm hos/cm at 25°c that means Na^+ is predominating factor. Water samples from Ajmer district show that EC value of 0.4 - 23.0 MM hos/cm with P^H value of 8.4 to 9.2, the water suffers from alkalinity as P^H value is more than 8.5, sodium ion. Here the predominating cation is Na^+ (Sodium) which varies between 2.8 - 1 0 Me/1.

But, the second criteria for determining quality of water based on the chemical properties, affects the growth and production of crops at length very much. Chemical properties of water are measured from the parameters of

1. SAR Composition and
2. P^H value of water.

While determining EC value of water, it is observed that salinity is caused primarily by Na^+, Ca^{t+}, or Mg^{t+} salts, and the cations play vital role. It has been found that the ratio between the three most important cations is best-expressed in the form of the so called Sodium Adsorption Ratio (SAR) for the purpose of assessing water quality for irrigations, is;

$$\text{e.g., SAR} = \frac{Na^+}{\sqrt{\dfrac{Ca^{++} + Mg^{++}}{2}}}$$

(by A. Bogenriender, Plant Research & Development, Vol.16, p.95)

Thus, the EC and SAR value correspond each other or rather dependent on one another.

But, the more effective measure of the quality of water is the P^H value which directly speaks of the acidity or alkalinity in soil, and makes the water qualities. It is the expression of Hydrogen concentration in water. It is observed that neutral water is that which has 1 part hydrogen cation concentration (P^H) per 10 Millions (10^{-7}) neutral solution. If hydrogen concentration increases the acidity also increases which is expressed in P^H value of less than 10^{-7} i.e., 6 or 5 etc. But, on decreasing hydrogen concentration the alkalinity increases which is expressed as higher P^H i.e., more than 10^{-7}. Thus, if the base elements of anions are dominating they will have the power to absorb more cations; Na, Ca or Mg, which in term will produce the effect in water solution and on decreasing hydrogen concentration (by way of evapo-transpiration) they produce effect of more salinisation or even alkalisation in soils. On higher concentration of hydrogen anions, the acidity produced by the single cation element (Na^t, Ca) is much more higher in ratio.

The effects of all these conditions of EC and P^H values of water, adversely hampers the growth of cultivated plants. The question arise whether the cultivated plants in serious cases, especially salt accumulating species can not tolerate the surplus salt in the soil. The best and in practice, the only method to achieve a saline balance is leaching. Na^+, Mg^{++}, and Ca^{++} ions move so freely in the soil and all therefore rapidly washed away by surplus water; only boron can lead to difficulties, which migrates slowly in the soil and has a toxic effect on plants with a danger of concentration at a place.

The maximum salt concentration that can be tolerated in the soil depends primarily upon the crop that is to be grown. Data-palms, barley, sugar beet, cotton and spinach are the species of salt tolerance. Whereas, citrus fruits, apples, appricots, peaches etc. are highly susceptible.

What can really be achieved as a saline balance in the soil for a given irrigation project depends upon the quality and the quantity of water. The effects of Electric conductivity on the yield of various crops may be seen in figure 7.

The yield reduction of crops due to EC value

The direct and the adverse effects of Na^+ (Sodium) the soil structure usually overlaps, with the result that in the individual case the respective contribution of these two factors to the fall in yield can be determined only with difficulty. The higher concentration of Na^+ in the soil is harmful to the sensitive cultivated plants. The clay minerals which are important for soil structure, are aggregates of silicate particles of varying composition and thus belong to the sheet silicates. They are a product of weathering and form a complex chemical process in the soil. These aggregates from the layers of SiO_2- Lattice, crystallised in the form of SiO_{-4} -tetrahedra with each oxygen atom, having side effects on the cultivated plants and adversely hampers the growth (Fig. 7).

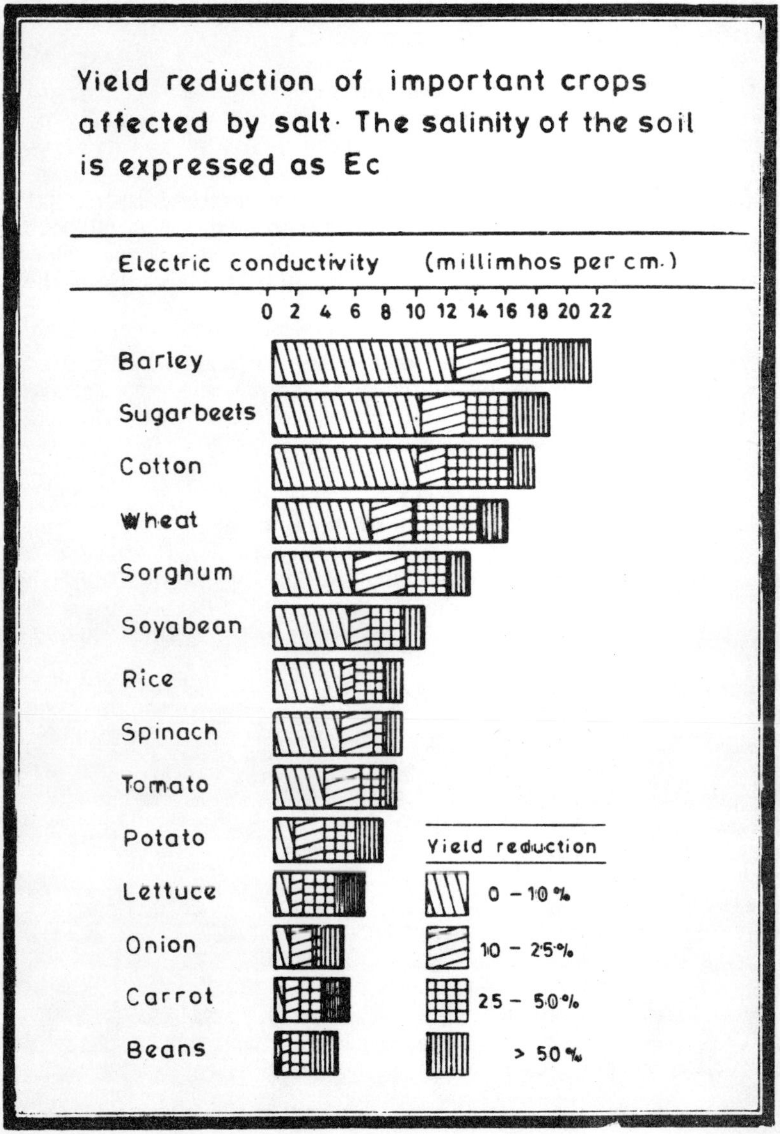

FIG 7

The state of Rajasthan is reckoned with varying degree of EC and P^H values in water, which is responsible for determining the quality of water, and its uses for cultivated plants. For the reasons, some sample studies, carried on in the desert are discussed here.

In Jaisalmer district, some observations at different selected places were obtained to measure the quality of water on the basis of EC and P^H values . The P^H value varying from 7.7 to 8.8 while the EC varies from 0.8 to 4.0 mm hos/cm at 25° C were observed. In all the water samples Na^+ was the predominating cation which ranges from 5.6 to 31.7 Meg/1 followed by Ca^{++} and Mg^{++} the soluable sodium percentage (SSP) ranges from 60.6 to 85.7% . All it shows that the water is not suitable for irrigation.

The Chemical analysis of well water in some of the districts were carried on by a group to study the P^H, EC and SAR values of the water and revealed some interesting figures which are listed below:-

TABLE-7

Chemical Analysis of Well Water

Dist.	P^H	Ec mmhos/cm at 25° C.	SAR (Na,Ca,Mg)	SSP	Dominating Cation/anion
Jodhpur (10 Samples)	7.3 to 8.5	0.9 to 18.0	3.70 to 125	3.7 to .45	Na^+ with Ca^+ Cl^-
Pali (10 Samples)	7.4 to 8.3	3 to 22.0	3.9 to 34.3	51.0 to 87.8	Na^+ Mg^{++} Cl^-
Bhilwara	7.1 to 9.2	0.6 to 16.0	1.8 to 50.0	65.0 to 95.0	Na^+/ Cl^-
Ajmer	8 to 9.2	1.6 to 23.0	3.7 to 69.4	92.6 to 92.6	Na^+ So^-_4 CI^-
Bharatpur	7.1 to 9.0	0.3 to 241.0	2.7 to 77.7	90.0 to above	Na^+ So^-_4

The recurrences of the above table reveal that the water problem at Jodhpur district is sever in many cases as it shows high contents of sodium (Na^+), high SAR, high conductivity and higher chloride (Cl^-) contents. This type of water is not useful for cultivated plants. The water at Pali found with high contents of sodium and chlorides but free from residual sodium carbonate ($Na_2 CO_3$). Bhilwara district also suffers from the saline water problems at underground as the residual sodium carbonates are found causing saline-alkaline with sodium contents. While at Ajmer the carbonates and bicarbonates causing the problem, with high contents of sodium and chlorides. Bharatpur district too, suffers from the problem of saline-alkaline soil, with the presence of higher sodium chloride and sodium sulphate in waters. SSP is higher due to sodium, and SAR due to the presence of sulphates.

Thus, in most of the parts of the state the quality of ground water varies from place to place that indicates the high contents of sodium and chloride alongwith high EC and p^H values. This type of water is not suitable for the cultivated plants and it makes the problem. Even in some other parts like Aravalli and Hadoti plateau regions, mixture of these bases making water unsuitable for the cultivated crops. But at places underground water is normal and can be used for crops.

It is very common feature that water is drained from the wells to irrigate the crops in the state. Therefore, the side effects of this type of water on crops are apparently not visualised but in long term their effects are noticed. The quality affects the crop yield in particular and production in general. The most affected crops are wheat, sugarcane, rice, citrus fruits and vegetables. Therefore, limited crops having resistant against salinity, arc cultivated. Jowar, Bajra, Gram, Maize are the crops which have developed salt tolerance capacity and therefore, cultivated in those areas where sodium sulphate or chloride contents are found excess the normal conditions in soil. In most of the Wn. parts of Rajasthan, the high Ec, p^H and SAR values have developed salinity or alkalinity in soil. So that, only salt tolerated crops are grown. It is also reported that in parts of Nagaur district, and in some other areas of Wn. Rajasthan, a new variety of wheat crop has been developed by the ICRI which has more tolerance power than the existing ones. But again it raises the question of disturbing the Eco-system of plant/soil generic relationship in general.

Gram and pulses are the crops which can withstand in such salinity affected areas and sustain growth in such conditions luxuriantly. But there is a problem of acute shortage of water at times in these areas.

Sources of Irrigation and Area Irrigated

Looking to the varied conditions of the water availability, under ground reservoir, and artificial storage and seasonal flow of water, various means of irrigation and practices have been developed in different parts of the state. The most common traditional sources of irrigation are found to be:-

1. Wells
2. Tanks and
3. Canals.

The state is very indifferently placed by sources of irrigation and the area irrigated by them. It is again found with unequal distribution of irrigated area as it varies according to the sources and their distributional pattern in the state.

It may be taken that wells are not uniformly distributed all over the state. Similarly, the number of tanks and canals are concentrated in areas of favourable conditions. Thus, if we look to the number of wells of all categories i.e., old and new ones, they are about 8 lakh including tube wells whereas , about 2.3 lakh wells are lying idle or without use. Well irrigation is very much practical in almost all the parts of the state as it has been an oldest and traditional source of irrigation. Even then, the well irrigation is very much associated with the areas of high water table, where lifting of water becomes an easy task for the farmers. Even then, the range of depth varies from place to place. In the Wn. Rajasthan it is observed that wells are found with more than sufficient depth of about 250 ft. whereas, in the eastern part it is only about 20 to 30 ft. The most favoured areas of well irrigation are Jaipur, Bharatpur, Pali, S.Madhopur, Udaipur, Chittor, Alwar, and Ajmer districts. While most scared areas are Bikaner, Jaisalmer, Churu, Sriganganagar, Jhunjhunu, and Banswara. As the construction of wells in the hard rock becomes a difficult task, and a costly, therefore in Hadoti plateau and elsewhere in the areas of hard rocks of Aravallis, well comparatively are not found in numbers. But in the Wn. Rajasthan the numbers of wells restricted due to the aridity and the under ground water table which go down to any unlimited depth. Even at times, if the wells are dug, it is not sure that the water which comes out will suit the crops, as in most cases it is found with high salinity.

In the areas of mountain and hills, some natural as well as artificial water storage in the form of Tank, are created. It is either due to the higher percentage of rain-fall or the natural landscape which provide the dumping grounds for water accumulation at first sight, and then developed into an artificial lake to serve both the purposes of natural scenery and water storage for irrigation of crops. Rajasthan has been on the brim of such affairs being princely states of Udaipur, Jaipur, Jhalawar, Dungarpur etc. and provided such developments. These places are well known for tanks and lakes, in this respect, it may be said that the former rulers had a good foresightedness for the future and created the beauty spots. It is just to meet the challenges of dryness and inadequate rainfall also. Thus, tank irrigation was also developed in vicinity of well irrigation simultaneously in these areas.

The area irrigated by sources is provided in the table below:-

TABLE-8

The Area Irrigated by Sources

Items	Irrigation Area in '000 l
1. Net Area Irrigated (Sources)	**8983**
a. Wells	1662
b. Tanks	112
c. Canals	941
d. Other Source	56
2. Area Irrigated More than Once	**766**
3. Gross Area Irrigated(Crop wise)	**3749**
a. Foodgrains	2551
b. Oilseeds	297
c. Cotton	317
d. Sugarcane	28
e. Others	476

The mechanism and technology of Irrigation by sources is discussed in the other part of the study, under Irrigation Mechanism.

SIX

Agriculture and Canal Irrigation

Though, the canal irrigation may not be termed as new, but the wide spread application of it in the mid-nineteenth century which has brought spectacular changes, may not be denied. The mechanism and technicality of canal irrigation has added new dimension only after the advent of introduction of innovative techniques.

The introduction of canal irrigation projects, with multi-purpose objectives, brought the revolution in the field of agriculture by adding water to the fields and luminating villages of the utmost distances. Canal system serves dual purposes viz :

1. Supplying water to the fields and
2. Controlling flood situation.

But, the difficulty of constructing a canal was felt at all levels and it became a national or state subject. Because an individual can not get construction of a canal and a state may not be granted full sovereignty over a river which is flowing in more than one state. Therefore, it is a National issue to 'resolve at a central level, besides state's interest in it. On national level three tier set up of irrigation projects were commissioned to utilise country's 57 million irrigation potentials at the beginning of the first five year plan-1951. The Big, medium and small irrigation projects, launched at National and state levels, trap the surface runoff of water and utilised for the crops.

Though, the working of major and medium irrigation schemes during the last three decades has been extremely unsatisfactory. Inordinate delay in completion of projects, raising cost, losses of water, surmounted unfavourable benefit and created an atmosphere of uncertainty at all levels.

Inspite of the poor show by big projects, irrigation mismanagement, financial burden, designs of large distributional system, impact on small farmers etc., are the glaring problems of irrigation schemes. Even then, their wide spread impact on the agricultural economy may not be ignored in the developing span of the new technology. The viability and feasibility of these projects during the period may be well visualised from the table below :

TABLE-9

Viability of Projects by Plan-Period

Plan period	Outlay/Exp. in Rs. crores			Irrigation potential cumulative (Mil-ha)		
	Major & Mid. Irri.	Minor Irri	Total	Major & Mid. Irri.	Minor Irri.	Total
Pre plan benefit	-	-	-	9.7	12.9	22.6
First plan	380	76	4[illegible]6	12.2	14.06	26.26
Second plan	380	142	522	14.3	14.7	29.09
Third plan	581	328	909	16.6	17.01	33.61
Fourth plan (1969-74)	1237	513	1750	20.7	23.5	44.20
Fifth plan (1974-78)	2442	631	3073	24.8	27.30	52.72
Sixth plan*	8448	1810	10258	-	-	-

* ET. Oct., 31, 1981, p.11.

India has one of the lowest crop yield in the world. The crop yield are erratic even in irrigated areas. It is estimated that irrigated land should yield 4 to 5 tonnes of grain per hectare per year. However, at present, it is hardly 1.7 tonnes on an average. Therefore, looking at the vast amount disbursed on major and medium sized projects, and the return yielded from agricultural fields, the importance of minor (small) irrigation projects or schemes, may not be ignored any longer. It is realised that the minor irrigation schemes may be more useful in boosting the states economy instead of major and medium projects.

Thus, in the sixth plan layout of the irrigation planning in the country, a major share of minor projects has been proposed to 1810 crore rupees, as against Rs. 8448 crores for major and medium. The additional benefits of this scheme is expected to 15 million hectares of land.

Under canal irrigation, the state of Rajasthan may be considered as fortunate in water reservoir and vast potentials of rivers and their catchment areas. But, all these resources are very poorly tamed and trapped so far. Some of important river basins and their catchment areas alongwith the land potentials are shown in the table below:

TABLE-10

River Basins, Catchment-Area & Land Potentials in Rajasthan

River basin	Total catchment Area in '000 Sq. Km.	Total Run off in Mil. $M.^3$	Cultivable land in '000 ha.
Chambal River	22.6	1,530	194.4
Banas River	47.6	1,280	342.4
Mahi River	22.7	1,600	60.4
Banganga River	6.6	96	-
Sabarmati River	2.7	150	-
Sukli River	1.0	255	-
Luni River	33.1	263	-
Gambheeri River	4.8	106	-

Besides these river basins, some other river systems have also been considered important from the minor-irrigation project point of view. Some of them have already been brought under construction of medium and minor projects. They are:

Berach, Kothari, Khari, Dhil, Kali, Sindhu, Mahi, Jakham and Anas.

The total area irrigated by Big, Medium and Minor projects in Rajasthan is given below:

The details of some of the important projects is provided here to have a comprehensive view of the canal particularly in the light of problems of regulating water to the cultivated crops.

TABLE-11

	Project	Area (in '000 ha)
1.	Bhakra Project (in Rajasthan)	3.02
2.	Chambal Project	2.11
3.	Gang Canal	2.82
4.	Indira Gandhi Canal (Former Rajasthan Canal)	3.56
5.	Medium Irrigation Projects	0.54
6.	Minor Irrigation Projects	2.28
	Total	14.33

Source: Statistical Abstract, 1981, Directorate of Eco. St., Raj., Jaipur, p.103.

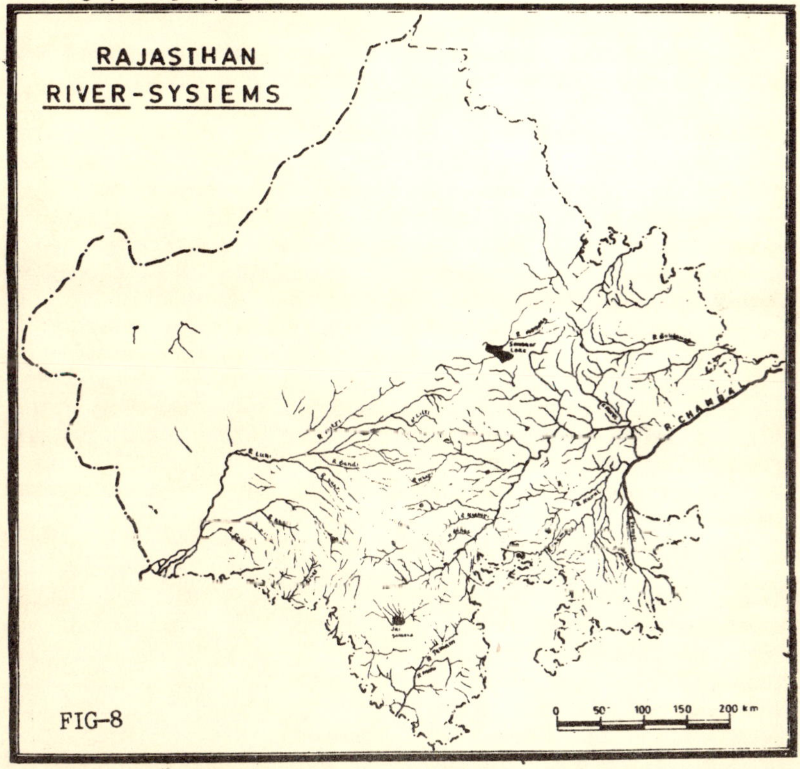

FIG-8

In Wn. Rajasthan, which is termed as desert, in respect of temperature and rainfall, scarcity of water is always there. To cope with the dryness, three big irrigation projects have been launched in the past three decades. They are :

1. Rajasthan Canal-now named Indira Gandhi Canal
2. Gang Canal and
3. Jawai Irrigation Project.

Indira Gandhi Canal - Project

This project proposes to transform an area of about 525 Kms long and 45 Km. wide stretches of the vast arid lands of Wn. Rajasthan. This project was initiated in 1948 by the then Chief Engineer of Bikaner, Shri Kanwar Sen, and then finally it was inaugurated by the then Home Minister Shri Govind Vallabh Pant on 31st March 1958. It was proposed to construct in two phases i.e.,

i Rajasthan feeder - 204 Km. (Fig.9) and
ii Rajasthan Canal -445 Km. but now it has been extended upto Gadra Road of Barmer district.

It was launched with the initial cost of Rs. 66 crores in 1958, but upto 1984 total cost has been incurred over Rs. 405 Crores with the incomplete work of the second phase. It runs nearly 9,425 Km. in length with 76 lakh acre feet of annual capacity of water in districts of Sriganganagar, Bikaner, Jaisalmer and Barmer with the total length of distributaries of about 2950 Kms., irrigating about 5.36 lakh ha. of land. It is estimated that it would be possible to raise at least 37 lakh tonnes of extra foodgrain annually.

With all the optimistic views, one may take the advantage of being a longest project scheme in the world. But the gloomy picture of the canal systems and the layout is before us, so far mechanism and technology of the irrigation system is concerned.

Basically, the area covered by such a big cannal is desert one, with high temperature and very low precipitation. So the aridity limit is higher. This tantamounts to dry conditions with highest rate of evapo-transpiration. Thus the losses of water through both the ways i.e., direct and indirect, is much more at all the times, than what has been estimated. Even then, the soils are not helping in preserving the moisture any longer. Even, if under ground water table is raised through additional irrigation water, the evaporation will absorb most of the water. It can not be preserved even by plants. (Pl. see aridity limit Fig.8) Secondly, the wide spread net work of distributaries also will help in evaporation of water, rather than protecting it.

Indira Gandhi Canal Project

FIG-9

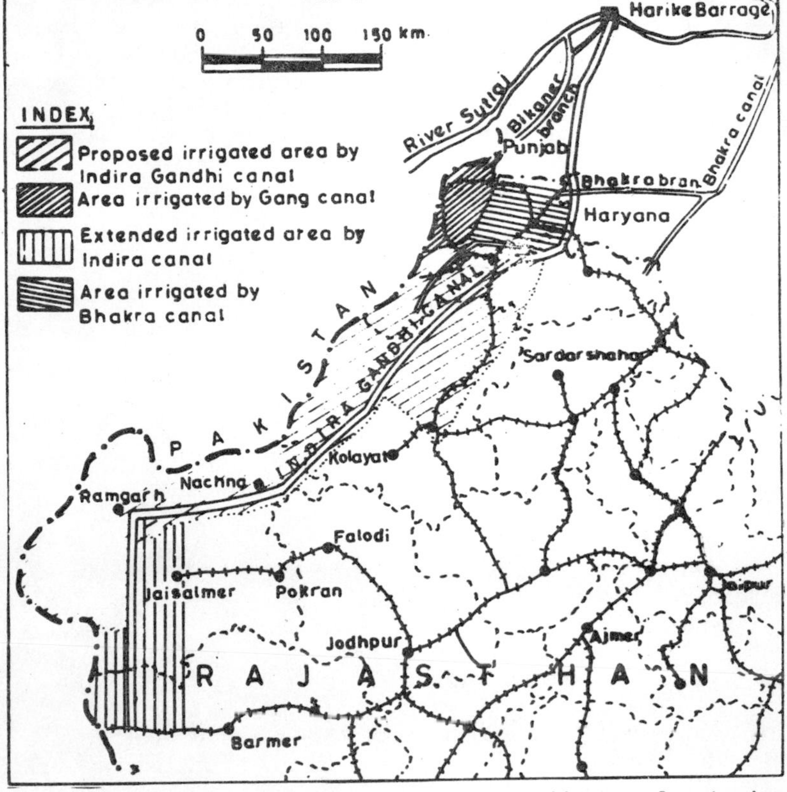

Thirdly, the magnitude of the side effects of watering the area is perhaps never realised by the planners, that how the water is going to react in the soils of high salinity or alkalinity. The base elements like KOH, NaOH, NH_4OH and CH_3COONa, $NaSO_3$ are lying in the beds of the sand. On percolation or leaching of water these contents rise to the upper layer alongwith the rising water level underneath. On evaporation these saline elements remain on the surface which may cause unaccountable harm to plants. The geological map of these areas speaks of such chunks or pockets where these bases, salts and lime contents are found abundantly at places. Further leaching process may lead to over-turning of chemical elements in the soil. The humas potash or sulphate contents applies to soils may go down on leaching.

Some of the apparent disadvantages of the unorganised and mismanagement of irrigation, have been highlighted by the International Environment and Development Institute. It stresses that the irrigation mismanagement leads to unnecessary loss of water, land erosion, water borne diseases, and use of more pesticides due to increase of pest and other bacteria. It has been envisaged in the report that those develop countries who ignore the general physical layout of the draining system and the water layout system, suffers more on account of such devastating havoc water impede or flood which damage crops and generate the better conditions for the bacterial congestion and many other insects. It is evidently stressed the fact that in 1970s the use of pesticides was 5 Kg. per person in the world, which rose to five percent every year in 1980s. The side effects and the reactions of these pesticides on about 5 lakh people were reported out of whom about 9,200 people died on account of the use of such pesticides.

On the other side, there were about 20 species who provided the resistance against the use of pesticides, 35 years ago at present their number have gone to 432, which are immuned to pesticides and surviving upon them.

The most striking account of these facts, has been provided by the experts committee on water on UN on the land development schemes in Pakistan. It says that about 100 ha cultivable land is turning into barren land every year. In Sahel of Africa, irrigation is provided to one ha. the other hectare is turned into waste.

Thus, there is no doubt that these incidences of land reforms may come to repeat the history in Indira Gandhi Canal also where scientists may have to fight for the coming menace later on.

Gang Canal

The another important canal in Wn. Rajasthan is the Gang Canal, the oldest in the state, constructed in 1927 by the then Maharaja of Bikaner Shri Ganga Singh. It is constructed on Satlaj River near Firozepur at Huseniwala, attains a length of 137 Km. from Firozepur to Shivapur with its total distributaries of 1280 Kms. The canal is wrecked at places being old of about 57 years. A link channel of 80 Km. is proposed to be constructed to join Indira Gandhi canal at Lohgarh in Haryana (Fig.10). The area irrigated by it is about 2144 cusec. The irrigation system known as "Barabandi" means turnwise is at work. It means water is released according to the fixed timings where no change is possible. In this system, some fields might be getting water at night and farmers have no choice to go for the another time even if they want. Thus, farmers have to go for only those crops, which may easily be irrigated at nights. Here it is a case of environment adaptation. Some examples may be cited from Sriganganagar, where Gang Canal command area receives water at different timings, particularly at night and day.

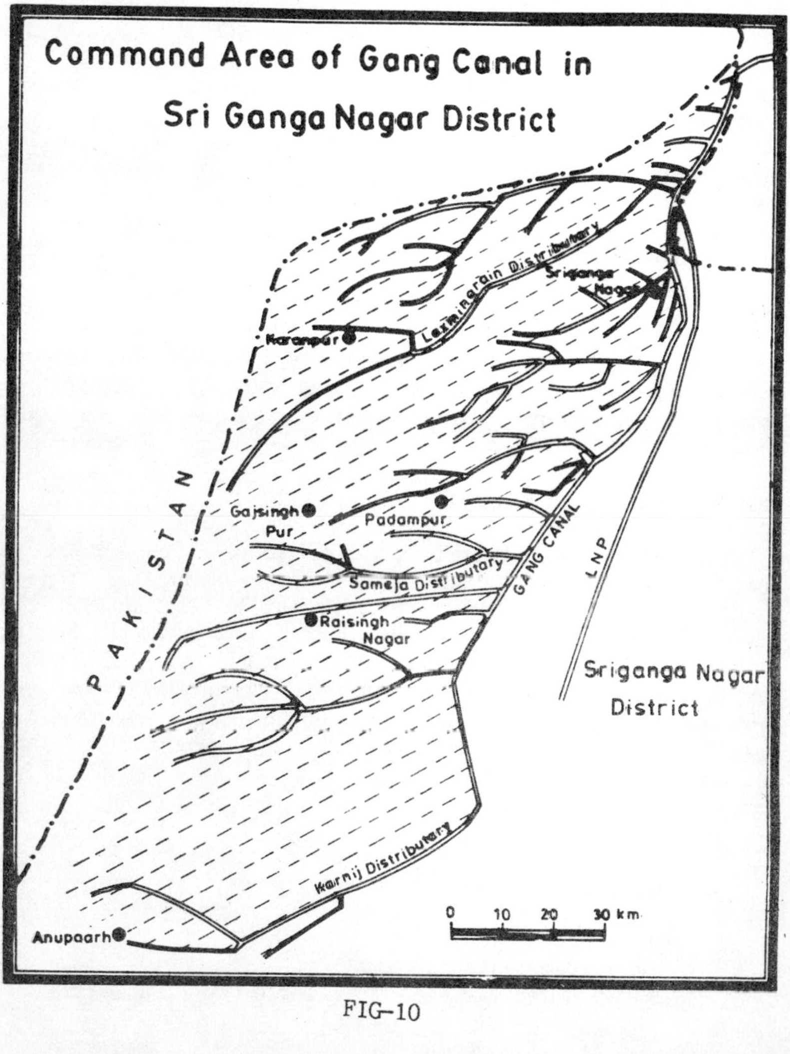

FIG–10

A field study was carried on to check whether there is any effect of Day and Night irrigation on crops; for this there was no option for farmers to supply water to the crops at their own will. So they have a set pattern of crops and their timings for watering them. In this process some fields were selected on the basis of Day & Night irrigation.

Further down-ward it joins Bikaner district it is a gift to Sriganganagar where a systematic irrigation is carried on. But, it is also a bone of contention for farmers, where water disputes are very common, which ends with killing of farmers. No doubt it serves the purposes of irrigation, but the cases of water losses are high due to mis-managed and now ill-fated one.

In south and SEn region of the state some important irrigation projects, both major and minor are under way of construction. Some of them are

1. Chambal project
2. Mahi project
3. Jakham project.

Some minor irrigation schemes are also under way of operation. They are Mezah dam, Khari dam, Gandhi dam, W. Banas project scheme, Orai scheme etc. are the important ones.

Chambal Irrigation Project

The second largest-irrigation project in the state is the Chambal Project. Actually it is a multipurpose project jointly envisaged by Madhya Pradesh and Rajasthan, on the river Chambal. The river chambal with its tributaries having a total drained area of about 1542 m. M^3, runs across the Madhya Pradesh and Rajasthan states. The project was initiated in 1943 with the construction of a hydro-electric station near Kota, later it was expanded to construct 3 dams shared by both the states in power as well as in water for irrigation.(Fig.11)

Thus, Chambal valley project came into being in 1953-54 with three hydro-electric project stations viz:

1. Gandhi Sagar dam
2. Rana Pratap Sagar dam and
3. Jawahar Sagar dam, later connected to Kota Barrage.

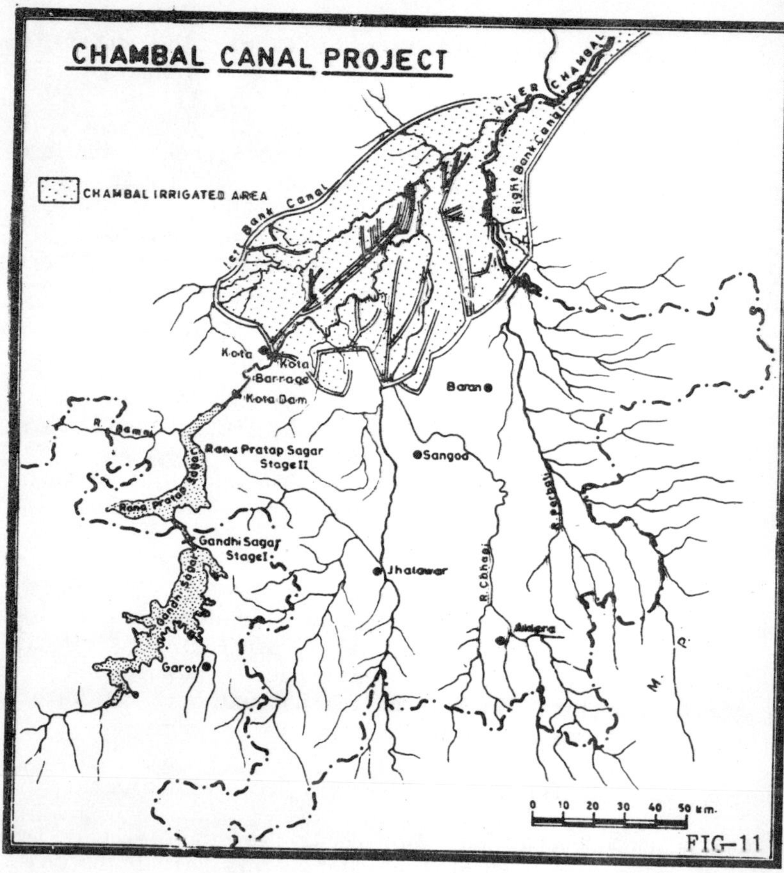

FIG-11

The Gandhi Sagar Dam was completed in 1960 with the net work of canal system in left canal and right canals.

The left canal runs through Bundi district and meets Jawai Ri. in Bhilwara, while right canal enters Madhya Pradesh after flowing about 261 Km. in Rajasthan. In Madhya Pradesh it has 641 Km. long course. The total irrigated area from these canals in both the states is estimated about 4.9 Lakh ha.

Rana Pratap Sagar Dam is constructed just 48 Km. north of Gandhi Sagar Dam in Rajasthan, forming another link of controlling water for hydro-electricity and to irrigate about 1.2 Lakh ha. of land.

Kota Barrage is an important water reservoir in a narrow escape to check turbulent flow of water. The water is utilised to irrigate about 5.6 Lakh ha. of land through 2350 Km. long and wide spread distributaries. (Fig.11)

Inspite of the best efforts by the engineers and the planners, the project in all its stages, suffers from some technical as well as mechanical defects. The nature's role by no why is less important in man-made creation. There are evidences of some of the glaring problems of water see page along the Kota Barrage which have been converted into permanent water logging areas.

These areas found parallel to the canal and have grabbed a sizeable cultivated land under bogs, which are continually fed by seepage from the canal. It is also reported that some base elements Sodium, (Na), Calcium, (Ca), Magnasium, (Mg) also have emerged on the surface due to over rising of water table in these bogs, when water evaporates in summer and thus, land has turned into a waste land.

But, the canal has opened the new vista of industrialisation in Kota, which may be considered a gift but on the contrary the industrial waste, particularly from the chemical industries (Shri Ram Fertilizers, Pesticides and Chemical Industries etc.) is adding pollution to the water, of which reactions are to be seen in time to come.

Mahi Bajaj Sagar Irrigation Project

In southern Rajasthan, Mahi Sagar Project is the second biggest scheme after Chambal project. It is said to be god's gift to the tribal people of Banswara and Dungarpur districts. The project has opened a new vista of the modern age to benefit the people of backward areas. This project was initiated in 1959-60. Since then, it is in the way of completion with all the ups and down as usually it is a nature of all such big projects. The total water capacity of reservoir is about 580 Mi.M^3 with the catchment area of about 6240 Sq. Km. It is proposed to construct a long main canal to irrigate about 30,750 ha. of land with the initial cost of about 182 crores of which 119.4 crores have already been invested by the end of March,1984.

With all the promising prospects of the Mahi Bajaj Sagar Project, reports have casted some gloomy picture and throbbing news reports under big head lines are attracting attention of all the people now a days.

Recently, the collector of the district (Banswara) has warned the people of the Mahi Sagar command area to vacate and shift to somewhere at a safe place, as the dam may be breched at any time due to over flow of water in rainy season. It is reported by the experts and engineers of the project that the project initially is based on soil based foundation instead of rockbed. This fact has already been established that the geological report about the basement (Underlying rocks) of the Dam-site, was wrongly surveyed and actual conditions were never known. Now, the basement of the dam, due to heavy load, is sinking down and creating new problems. Because, the underlying strata are unable to hold the load.

There are also reports that newly constructed 72 Km. long canal is leaking and seeping at places due to faulty construction and thin cement plasterings.

In vicinity to the Mahi-Sagar Project, some new irrigation projects in southern Rajasthan were started in 1960s. Amongst these projects, Jakham, and Menja Dam are important ones. Jakham project was initiated in 1962, to irrigate the area of Chittorgarh, Dungarpur Tahsils. It is still under construction with the hope that it will provide the water to 21 thousand hectares of land on completion. Similarly, Menja Dam was constructed in 1957, on Kothai River, 8 Km. away from Bhilwara, to facilitate about 10.5 thousand hectares of land.

A series of irrigation scheme have been initiated at various places like Khari in Asind (Bhilwara), Adwan Dam in Shahpura (Bhilwara), Orai in Chittorgarh, West Banas project in Pindwara and Gambhiri in Sawai Madhopur.

In SE and eastern Rajasthan a number of small and medium irrigation schemes have been commissioned and some of them are operating well. The well known irrigation projects are Morei Parbati, Bilas irrigation scheme, and Pipalda lift irrigation scheme, Indira lift irrigation channel and Gambhiri Scheme are among the longest channel schemes, through which Sawai Madhopur, Chittorgarh and Bharatpur districts are linked respectively.

Besides these irrigation projects there are about 50 medium and 614 small schemes have already been started in the state. Thus, the state has 2 big multipurpose projects,

5 big, 14 medium and 20 small schemes which are under way of modernisation. The expected water potential in the state for the further irrigation planning particularly in the Wn. parts, are indicated in the table below:-

TABLE-12

Irrigation Planning Perspective.

District	Yearly Capacity of Refilling in Mi.M^3	Yearly Prog-gress in Mi.M^3	Additional Resources in Mi.M^3
1. Sriganganagar	155	94	61
2. Churu	210	13	197
3. Jhunjhunu	309	102	207
4. Sikar	452	161	291
5. Nagaur	545	135	470
6. Bikaner	91	8	83
7. Jodhpur	445	145	300
8. Barmer	193	64	129
9. Pali	437	182	255
10. Jalore	445	121	324

Source: Rajasthan, 1984-85, Rajasthan Sandurbh Granth, p. 155-157.

SEVEN

Irrigation Mechanism and Technology

Irrigation by various sources like wells, canals and tanks etc. has been a traditional practice in different parts of the earth and still dominated by the old methods and techniques of irrigating fields as they were developed in the past. Thus, farmers have learned by experiences through past centuries that how and when to irrigate the fields. In this context, many ways and means of practicing cultivation have become obsolete but irrigation practices still dominate the scene with old similar fashion.

In the present day context, it becomes a point of investigation that old practices need changes and wider perspective in exploiting the natural resources. For this reason, irrigation needs scientific explanation of its mechanism and developing techniques in developed and under developed countries, where it too, has become a subject of rapid change in terms of :

i) Soil erosion
ii) Requirement of adequate water for crops
iii) Water management - Preservation of water from losses and
iv) Optimum use of water for irrigation.

These have been the most pertaining questions before the irrigation engineers and planning experts engaged in the development of agricultural extension plans. Thus, the irrigation mechanism and techniques have become the most important aspect of the geographical explanation too. Because, irrigation mechanism and techniques are equally associated with the traditional social systems, organisational practices and production capacity of the land. The maximum use of potential water resources is the need of present day, which is possible only when the size of holdings are made ideal, field parcels are adequately spaced, and water channels are thoroughly regularised. To prevent losses of water through various means, the scientific knowledge of soil and terrain is an essential aspect which not only determines the regulation of water but also control the moisture, water balance and water regime in soil. All these factors affect irrigation in a long term, but altogether change the over all characteristics of the cultivation. The consequences of unscientific irrigation are evidently seen in many parts of the country, where over all production is adversely hampered. Excess or inadequate water for crops is not an ideal condition. This point draws attention of the geographers and all other planners, the world over, specially those who are engaged in planning of all those critical areas falling in danger of ill managed practices of irrigation.

Mechanised Irrigation and Technology

Irrigation mechanism and technology are the usage of the modern development, associated with the span of irrigation with the pace of time in developing countries. Irrigation mechanism is the process through which water is released in order to protect soil depletion, atmospheric evaporation and proper utilisation of maximum water for crops. Whereas, irrigation technology deals with methods through which one makes the best use of water potentials in order to obtain more production from the fields.

Thus the application of water to the fields means to :

i) Provide adequate water to the crops in time ; and
ii) Prevent water from ground and atmospheric losses.

Water losses are reported heavily due to seepage and evaporation processes. Therefore, the problem of regulating the water through a systematic way by various means i.e., canal, well and tanks, is a question of paramount importance in irrigation engineering, and thus, irrigation mechanism becomes an important part of such studies.

The traditional means of irrigation is being practiced since the dawn of human civilization and no change has come to the fore, even in the modern phase of human skill and developed technology. In contrast with the developing technology of agricultural engineering it is being considered a vital problem. Because, cropping patterns, variety of crops, high yielding variety seeds, soils fertility and amelioration efforts of soils, etc. have been developed to a considerable level. But the water regulation system is still captivating usual line of field techniques.

The Conventional Methods of Farm Irrigation

The age old traditional methods of farm irrigation are the Canal, Lift and Basin irrigation. These methods are found in conventionalised manner all over the world, whether it is oriental or occidental. There are two techniques of lifting water to irrigate fields viz; one, by charas and second by rehant (Persian Wheel). A third method recently developed is known as water sprinkling method, (tube or pipe irrigation). Under first system of irrigation, much manual labour is involved with a sufficient time; i.e., 17 to 28 hrs for 1 ha. of land. But it again differs from source to source and means to means.

It is observed that where Charas as a means of irrigation is used, it takes much more labour and time as compared to Rehant, the persian wheel. But both are traditional means and less costlier. Charas is easily managed by a farmer or a house-hold. It is used in drier or semi-drier parts where water is not abundantly available with a regular underground flow. The another determining factor is the depth of water table which again vary from place to place. The load of water after a sufficient depth can not be lifted. Therefore, water is lifted by Rehant from the ground floor where it is abundantly available.

The technique to irrigate fields deployed in both the methods is common, i.e., through strip fencing formed around the parcels. But, the usual loss of water is unavoidable.

It is noticed and observed on the spot that charas require more manual labour and time, hence water losses are 2 times greater than canal irrigation. Because, the flow of water is not frequent and constant with equal amount of water in the strips. It causes evaporation as well as seepage and absorption by soil in a given time. While canal makes constant flow with a limited requirement of water in a given time. Water covers more space and distance in a regular flow.

The effect of these systems can invariably be seen on the production of the crops also. Irrigated wheat with Rehant gives higher yield while, irrigated wheat by charas reduces the yield.

Canal irrigation system is quite different from both the above mentioned techniques. Though canal water is made available to flow in channels and then to sub-channels from which water goes to the strips or furrows. Thus, it travels a long distance over miles together. It is the most widely practiced system of irrigation in most of the developed and under-developed countries.

Canals are generally constructed and organised by the Central Government agencies. Therefore individual farmer is no more concerned with the planning and layout of the canals. He (farmer) is only concerned with the water which he gets from canal irrespective of the time. Government officials are assigned work to release water according to the time-table. They mark the Zonal areas and their timing with the amount of water to be released. Thus, farmers are living on the mercy of these officials that at what time they allow water to flow in or let the areas be dry on.

Here, it would be appropriate to mention that day and night irrigation makes an important point in farm mechanisation. This point would be elaborated at a later stage, but needs thorough study exclusively on the effects of day and night irrigation on crops.

The conventional methods of irrigation lead to many untraceable problems and causes water to flow in abated conditions leading to great losses of potential reservoir. Irrigation water is lost in many ways, important among them are:

i) Impeded drainage lines to the fields;
ii) Seepage of water;
iii Over or under watering of crops;
iv) Evaporation from fields;
v) Drainage leakages; and
vi) Day and night irrigation and its consequences.

Requirement of the Mechanised Farm Irrigation

Mechanisation of farming in order to regulate water to flow in or allow the water to the fields more quickly, promptly, rationally and better than with traditional simple methods, require a more effective means and well developed irrigation technology. The following aspects may be considered important in order to achieve trouble free mechanisation of farming in irrigated areas ;

1. Arrangement of the fields;
2. Requirement according to the state of the soil
3. Requirement according to the irrigation system.

1. Arrangement of fields in respect of -

i) Unimpeded access to the field; and
ii) Convenient arrangement of the parcels.

i) The water should have impede-less channels to flow in so that it may reach to the opposite end within a specified time with a required amount of water for crops to be retained in soil. Unmanaged field arrangements may lead to over watering or under-watering conditions which may adversely affect the growth of crops. The amount of water reaches to the soil depends upon the rate and speed at which it flows in a channel. It is also checked by soils. The track system itself should be laid out in such a way that it can be travelled quickly and that the optimum load can be transported on each strip.

ii) The convenient arrangement of the parcels depends upon their size, shape and the length. Actually it depends upon the size of the holdings and their pieces. The parcels are made-out according to the size of the fields. So that the fields may efficiently be managed for work with a shortest possible time. The actual working time available to farmer is about 3-1/2 to 4 hours, in whole day. Thus, the size and shape of parcels is an important factor in field arrangement.

The square shape of parcels is generally regarded as ideal one in many respects. It is the most common shape practiced all over the world, but it depends upon the size and length of the parcel even, since it is hardly possible on small and medium sized holdings. If the length of parcel is greater, than the parcels are made rectangular.

2. Requirement according to the state of the Soil -

Soil is the most crucial factor in determining the growth and overall production of a plant. Soils play determinant role in shaping regions economy, as they behave indifferently according to their composition, structure , and characteristics. So the water requirement also differs as per the types of the soil i.e., alluvial, loamy, sandy, and black etc. The moisture retaining capacity is greater in black soil while lesser in sandy.

Similarly, different types of soils have different water regime, air regime, soil temperature and moisture capacity, which is available to plants in their growth period. How different types of soil function with the available water (irrigation, practices) in a flood basin area, is a complicated process.

3. Irrigation system

Irrigation system means all the components and equipment required for irrigation and also for supplying the water and distributing it over the fields. This requirement depends upon the water supply and distribution of water in the fields. The capacity to irrigate all the fields, with sufficient amount of water and with great deal to even supply of water to each parcel is a factor of efficient and skilled irrigation. This efficiency will keep the water losses to the unavoidable minimum amount.

The traditional systems are well marked with the means and potential source of water in any region. Since long, wells, canals, and tanks have been the major and only means of irrigation not only in traditional societies, but also in modern technologically developed societies. Therefore, irrigation system is closely associated with lifting water and channels (draining). These have been the conventional methods of lifting water and take to the fields. In contrast with these techniques and systems, German Engineers have developed two different types of technologies suited particularly to the desert regions, viz; Trench irrigation and Tube irrigation. However, Basin irrigation and flood, plain irrigation systems have been the techniques of the old world where unmanageable flood water causes havoc to the extensive low lying areas. But, the developed system of irrigation needs more emphasis to prevent losses of water in order to optimum utilisation.

While dealing with irrigation mechanism & technology two aspects are considered more important to be discussed here. They are:

1. the field arrangements (Parcels) ; and
2. day and night irrigation.

1. THE FIELD ARRANGEMENTS

As it has been emphasised earlier in the preceding pages that the arrangement of the fields, and field systems affects water in many ways. It depends upon the field arrangements that how the water is to flow and at what speed. Therefore, to let the water flow more freely and smoothly, field parcels must adequately be arranged according to the nature of relief and the character of the soils. Parcels vary according to relief and terrain.

Types of the Parcels

According to the size of the holdings and layout of the fields, parcels are suitably spaced in between water strips or furrows connecting each parcel at the opposite end. In field mechanisation it is necessary to regulate water through parcels in a required time, so that adequate water may be available to each parcel in a given time. Since the shape and size of the parcels are chosen as squared or Rectangular, according to the suitability or terrain and slope of the land. The shape and size of the parcels affect the flow of water in terms of time and amount of water to the soil, which in turn affect the quality of crops and thus production. This fact may be summerised from the study of Prof. Werner Doppler's observation 'irrigation as basis for development'[1] that water is utilized differently by the plant during the different growth phases. Yield experiments with rice as a function of the time at which a physiological water deficit occurs produced the following results in the Phillipines.

Adequate water supply	6.2 t/ha	8.1 Kg/mm Water.
No water from 41st - 81st day	4.4 t/ha	5.6 Kg/mm Water.
No water from 63rd - 102nd day	2.0 t/ha	2.5 Kg/mm Water.
No water from 63 days to harvest	0.6 t/ha	1.5 Kg/mm Water.

If we consider ideal shape and length of the field, for irrigation, ignoring the other factors (ownership conditions) the basin is determined by irrigation methods to ensure adequate efficiency of the water distribution, the size of the individual flood basin is limited in keeping with the type of soil and the available water, as the em-

---

1. Werner Doppler, Plant Research and Development, Vol. 17, 1983, Institute for Scientific Cooperation, Tubingen, W.Germany, pp. 26-36.

pirical values collected by Booher (Peter Wolff).[2] It is shown in the table below :

TABLE-13

Suitable Basin Sizes for Flood Irrigation as a Function of the Soil Types and the Available Water.[3]

WATER INPUT 1/s.	M^3/h	SAND	PADDYFIELD SOIL	MEDIUM	CLAY
		Basin size (ha)			
30	108	0.02	0.06	0.12	0.2
120	432	0.08	o.24	0.48	0.8
210	756	0.14	0.42	0.84	1.4
300	1080	0.20	0.60	1.20	2.0

It is evidently clear from the table above that the size of the basin in accordance with the available water varies from soil to soil and amount of water. Thus the operation of fields may adequately be carried on with 300 M as optimum length and 30 M width. The suitable basin sides may be considered as 1:1 to 1:4 M. in width and length as ideal one.

In Indian conditions this type of layout of a basin and shape & size of a parcel is not possible in all cases. Though the most common shape of the parcel is found as Rectangular and in other cases square one. In hill tracks and in narrow strip of fields triangular shape is preferred. In foot-hill region parcels are squeesed to triangular form but generally it depends upon the shape of the terrain.

What so-ever may be the shape and size of the parcels it affects the irrigation. It is observed that in Rectangular shape of parcel (Fig. 12A) water does not move out swiftly. If we take the size of a parcel in ratio of 1:1 M, then water takes a little more time to travel from A to B point as compared with rectangular of 1:2. But it again varies from soil to soil. Thus, the irrigation as a function of

---

2. Peter Wolff, "Mechanisation in irrigation farming." Applied Geography and development, Vol.19, Institute for Scientific Cooperation, 1983, W.Germany, p.10.

3. Booher, L.J. Surface Irrigation - F.A.O. Agricultural Development, Paper No. 95, F.A.O. Publication, Rome, 1974.

soil types is clearly determined by the process of irrigation technology. If the parcels are aliened rectangularly, the water takes a little time to travel from A to B point, preferably be of 1:4 as ideal size (Fig.12B). Because, in rectangular form water moves freely with a soft-speed in parcel rather than in squared size of parcel. It is because of the number of parcels and the number of the strips through which water travels. In square size the number of parcels and strops increases while in rectangular it is limited. Other factor for water movement is the sprawl of land parcels. If they are squared then the water will spread gradually in all directions first and then it will move forward. Thus, in the shape of square, water will take 10 to 12 minutes to cover or to fill the parcel if it is being irrigated by charas as a mean. And if the water is supplied by the pump set then it will cover just double the area of a parcel by charas. This is also a factor of depth of the water level in the well. Whereas in rectangular size, water moves onward by striking with the sides of the parcel which are narrower. Thus, in such parcels water takes about 8 to 10 minutes to fill it up. If the whole process in irrigating one hectare of agricultural land is thus taken into account, then in rectangular formed parcels with charas irrigation, it takes about 17 hrs, while in squared formed parcels about 26-27 hrs. It is also observed that in squared parcels water losses are greater on account of the number of strips and parcels as well.

To form parcels is a very technical point with which farmers are very well acquainted. In a 6° to 10° sloppy field areas straight lines are rarely preferred in parcels. They are generally of slanting lines inclining towards lower side of the slope and then parcels are arranged in a rectangular form (Fig.12D). It is reported that if parcels are arranged in a straight lines, the last point of the parcels on top side (shown by X Fig.1E) will take time to fetch water and thus the parcel will be over loaded by water in rest of the time. On the sloppy terrain, as it may be the case, strips are marked along the parcels aligned with the slope of the land.

Triangular shaped parcels are only preferred in a narrower strip of the land or squeesed margins of the fields at hill side or any such feature which bound its borders. Normally this shape is chosen in hill tracked areas with the narrow strip irrigation. But as a mechanism of irrigation is concerned it takes a quicker time to sweep water from one end to the other, and thus soil remained unquenched or moistureless.

WATER REGULATORY SYSTEM

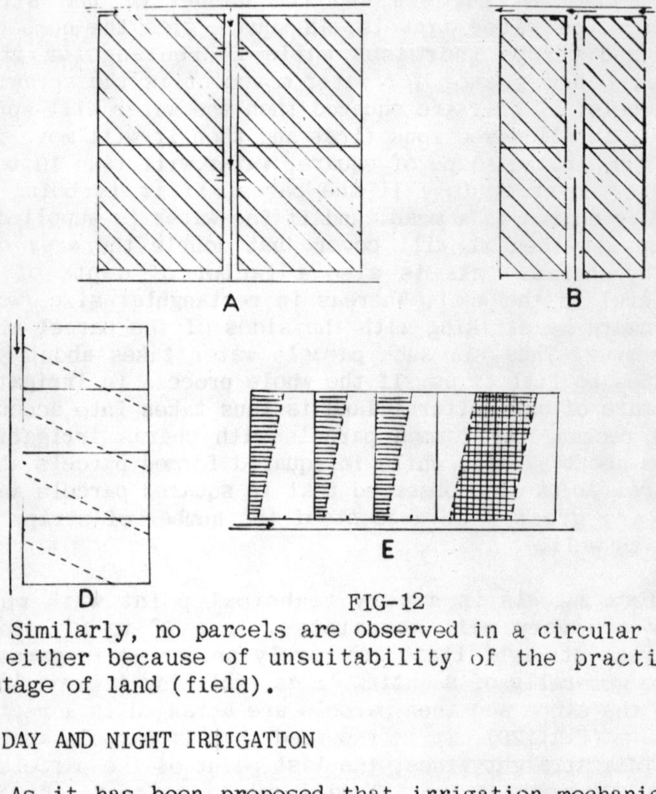

FIG-12

Similarly, no parcels are observed in a circular form it is either because of unsuitability of the practice and wastage of land (field).

2. DAY AND NIGHT IRRIGATION

As it has been proposed that irrigation mechanism is a function of soil types and the process involved in is rather technical one. Therefore, time is the important factor in such development efforts and irrigation planning.

Plants and certain crops receive water in the day while some in the night. At night, crops accumulate nitrogen and other trace elements to complete micro-organism circle for photosynthesis. They react with the hydrolysis process alongwith chemical and biological processes and proceeds for the development in accordance with the conditions of ash-food and sustain growth upon the day light-heat. This process is fixed according to the availability of water in set timings. If the watering process is reversed in such a critical moment the whole process is adversely affected in reducing crop yield and production.

In some cases crops are watered in the day time. But, the incidences of water losses through evaporation and even transpiration are greater. So the water for plants to accomplish Bio-Chemical process is not available and thus, micro-organic cycle remains unfinished. Thus, plants die or go on wilting in absence of adequate amount of water. Water helps the crops in their nitrogen fixation and metabolic processes. The efect of water on soil is seasonal and so on the plants. The temperature of soil is seasonal and so on the plants. The temperature of soil is directly affected by the amount of water in the soil. Thus the adequate amount of water regulates the soil metabolism (Micro-organic substances) which is sustained upon crops by various processes in the presence of sunlight, heat and temperature at the day and night. Thus, day and night irrigation makes an important aspect of the study as crucial factor. Some more effects of day and night irrigation on crops are highlighted in unit eight.

Therefore, mechanisation of farming in irrigated areas is a complex subject of the constraints imposed by the nature as well as man, specific to the prevalent irrigation systems and the access of the field patterns. Answer can be provided only when the system of farm tracks be ensured technically of high level with adequate water supply to the crops.

Besides all the technological efforts made in order to develop mechanised irrigation farming, a few points may be of worth mentioned.

i) Nitrogen fixation and losses;
ii) Application of fertilisers (NPK)
iii) Green manure is applied ; and
iv) Water strips or furrow' width.

The crucial amongst the problems is the regional disparity in supply of water, with indifferent timings. Decisions for supplying water to the crops are taken at the political level with biased attitude. This partial view leads to release water with insufficient amount either in the day or at the night. It seems logical to find a rational and most reasonable solution to the problem in a common and acceptable integrated national policy of irrigation system.

Irrespective of the development of big irrigation projects (Dams) small scale projects may serve the greater cause of farmers in tapping source of water in small running units. This may reduce the cost as well as the tension of large proportion of the farming class.

EIGHT

The Conditions of Aridity Limit and Humidity for Cultivated Plants

INTRODUCTION

It is the water which makes the matter more crucial for the growth and development of plants. In fact, requirement of water varies from plant to plant and in time to time. Too much or deficit amount of water for plants in accordance with the surplus or deficit hydrological budget, hampers the plant's generic process and does not allow plants to nourish and flourish.

For the survival and generic process of plants aridity and humidity are the essential conditions, but, involves many complexities of meteorological observations and calculations. The extent of aridity and range of humidity for the physiology of cultivated plants occurs very frequently in countries facing seasonal variations and changes of temperature and precipitation. Changes in soil moisture directly related to temperature and precipitation, further complicate the process of water regulation to plants.

The water regulatory channels, on the contrary, behaves indifferently in the areas of too much aridity or too much humidity. Thus, aridity and humidity are two prerequisite conditions before, one proceeds with the watering the crops. For the proper utilization of water to the plants, aridity limits and conditions of humidity are the true determinants, through which not only the requirement of water for crops can be met but also the timings may be adjusted.

When and where water is required by plants and to what extent are the problems of paramount importance. Aridity and humidity conditions may limit these problems and may pave the way for the better regulatory systems to achieve optimum conditions of the watering the plants. Therefore, in the present study attempt is made on two important factors; determinants of (i) aridity limit and (ii) humidity conditions, varying with mean monthly temperature and precipitation, in the state of Rajasthan, (Applied Geog. & Development, Vol.17, 1981) by applying D. Schreiber and W. Schiedecken's methods, with the objectives of:

(i) To identify and demarcate the limits of aridity and humidity conditions in the state, and

(ii) To locate areas of irrigation its feasibility and application.

ARIDITY LIMITS

Penman (1948) was the first to introduce the measures of aridity limits, but was without much concern to attitude as a factor for adjustment. Later Jatzold (1976) used Penman's equation with altitude adjustment and multiplied by the factor 0.4 for the monthly aridity limit. But, all these values were not competent enough to give satisfactory results of aridity limits, as air pressure according to altitude and the sun shine duration cause much concern to precipitation and evapo-transpiration process. Thus, it is a problem to obtain precise meteorological and climatological observation to compute air pressure according to altitude and length of sun light, in most of the developing countries like India.

However, for the reasons, a technique employed by D. Schreiber (1973) to measure aridity limit with certain correction and adjustments of the records of the meteorological stations for the mean monthly in °C and precipitation in mm., found to be more suitable for the present context. The present method undertakes the natural physical and plant physiological conditions for the evapo-transpiration process into consideration. This technique may help in identifying and demarcation of Agro-climatic regions.

Temperature, precipitation and altitude all are correlated factors which affects each other and therefore, air pressure and duration of sun shine changes according to different locations. It may be assumed that the evaporation rate increases as the air pressure decreases i.e., as the altitude above sea level increases, the duration of the plant transpiration each day changes as a factor determined by the length of the sun light, soil temperature and by the landscape covered with vegetation. All these factors combined together exert influences on the rate of evapo-transpiration by one way or the other.

Schreiber took all these factors together and prepared tables showing monthly aridity limits in mm of rainfall with corrected mean monthly temperature, (tk), in one and additional temperature allow for the altitude factor when calculating aridity limit in the second. He also prepared a third table to show the length of sun light (day light) with addition and substraction for the exect evapotranspiration. All these tables are corrected and adjusted in such a way that they may be used for the Northern hemisphere to obtain aridity limit (Schreiber, Applied Geog. & Development, Vol 17, 1981, Berlin).

The corrected temperature employed in the method, increased by the altitude factor (Δt_h) so the mean temperature is t = tk near sea level. The added altitude factor t_h, increases with both the altitude and the mean monthly temperature. Thus, the correction and calculation of t+ t_h = tk may easily be made (Schreiber, 1981).

On the basis of the above suggested methodology and from the determinants of the aridity limit, some critical results are obtained while calculating aridity limit for the state of Rajasthan. For the application aspect, four units as representative of different physiography of the State have been undertaken. The Western desertic conditions are well represented by Jodhpur district., while in the south-east, Kota represents the wet conditions of the higher precipitation belt. In the central part of the State some examples have also been drawn and tested from Ajmer, Udaipur and Jaipur districts.

Characteristically, all these units represents not only different physiography but also marked seasonal variations of temperature and rainfall, making irrigation as an essential part of agriculture. Aridity limit in all units is distinctively marked to the extent of necessitating irrigation and the range of its feasibility in the state as a whole. The well marked aridity limit and humidity conditions or water balance in soil are shown in Fig.13 for which analytical explanations are provided below:

JODHPUR DISTRICT

Jodhpur district lying in the western part of Rajasthan climatically falls in the desertic fold of the 'Thar Desert' and characteristically it is over-ridden with all those con-

TABLE-14

JODHPUR DISTRICT, (26°-20°N,73°0° E, 264M) Desert Region

	J	F	M	A	M	J	J	A	S	O	N	D	
	16.8	19.2	24.7	29.5	33.5	33.8	31.3	29.1	29.1	28.9	21.9	18.2	Ct
	1.0	1.0	1.5	1.8	2.0	2.0	1.9	1.8	1.8	1.6	1.4	1.0	th
	17.9	20.2	26.2	31.3	35.5	35.8	32.2	30.9	30.9	28.5	23.3	19.2	t_k

From Table I, it give the monthly aridity limit of

46	53	73	92	109	110	95	90	90	81	63	50	mm

The period of day light is obtained by deducting or adding from table-III

-2	-2	+1	+2	+3	+3	+2	+2	+2	+1	-1	-1

So the precipitation is sufficient only when it is over

44	51	74	94	112	113	98	92	92	82	62	48	mm r

Actual mean monthly precipitation

	3.7	6.7	2.7	3.2	10.25	35.5	99.2	122.0	60.0	8.0	2.7	2.7	mm
p	0.37	0.67	0.27	0.32	1.00	3.55	9.9	12.2	6.0	0.80	0.27	0.27	cm
E_t	8.0	9.1	11.5	13.9	15.9	16.0	14.8	13.7	13.8	12.7	10.4	8.6	cm

ditions which dominate desert or semi-desert, and thus, provides the best example of the aridity limit, which is calculated in the table below (Though, Jaisalmer district could have been the true representative unit of THAR, but due to unavailability of evapo-Transpiration data, Jodhpur has been taken).

On the basis of the above recordings one may deduce that there is a wide range between actual monthly precipitation and the calculated aridity limit in mm, except two incidences of July and August (Fig.13-A). These two months may be considered wet months for the whole period otherwise, rest of the months experiences dryness. The period of dryness spreaded so extensively that the month of July almost comes on the margins of aridity limit with a little difference of 99.2 -98 = 1.2 mm . During this period sun is on the peak, reduces the humidity in soil. The effect of aridity may further be explained from Fig.13B, which is drawn to visualise the actual water balance or budget in the area.

The above figures (P & PEt) for the water balance in Jodhpur district (Fig.13-B) reveal that there is no any relationship between precipitation line and evapo-transpiration line as they do not meet on any point for the entire period of 12 months. There is a negative correlation between high temperature and the low humidity. It is evidently clear from the figure that there is a very high evapo-transpiration while precipitation received by this unit in the month of July and August i.e., 9.9 and 12.2 cms. is not enough to cope with the high rate of evapo-transpiration of 14.8 and 13.7 cms respectively, hence there is no water balance. Soil go without moisture in almost all the 12 months.

These conditions have ceased plant growth and have caused a grave concern for rising limit further. In absence of humidity in the air and moisture in the soil plants are perished only to the limited sources of underground water or the wells. Irrigation in such areas becomes more difficult as water swiftly evaporates or even transpirates from plants. Therefore, only drought resisting (xerophyting) plants are surviving.

KOTA DISTRICT (25° 15° N, 75° 45° E, 274M) Plateau Region

In contrast to the desert region, in the west, Kota provides the incidences of wet conditions in the SEn part of

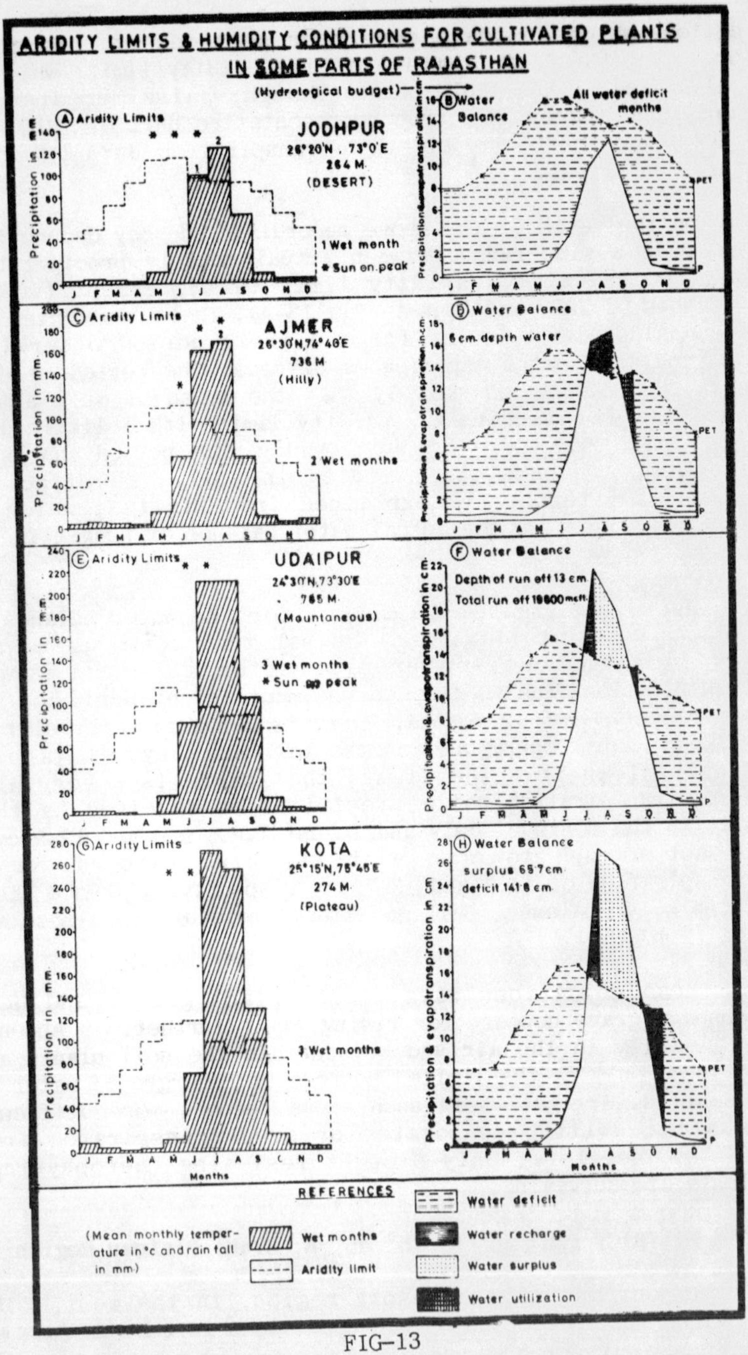
ARIDITY LIMITS & HUMIDITY CONDITIONS FOR CULTIVATED PLANTS IN SOME PARTS OF RAJASTHAN
(Hydrological budget)
A Aridity Limits
JODHPUR
26°20'N, 73°0'E
264 M
(DESERT)
1 Wet month
* Sun on peak
B Water Balance
All water deficit months
PET
P
C Aridity Limits
AJMER
26°30'N, 74°40'E
736 M
(Hilly)
2 Wet months
D Water Balance
6 cm. depth water
E Aridity Limits
UDAIPUR
24°30'N, 73°30'E
765 M
(Mountaneous)
3 Wet months
* Sun on peak
F Water Balance
Depth of run off 13 cm
G Aridity Limits
KOTA
25°15'N, 75°45'E
274 M
(Plateau)
3 Wet months
H Water Balance
surplus 65.7 cm
deficit 141.8 cm
Precipitation in mm.
Precipitation & evapotranspiration in cm
Months
REFERENCES
(Mean monthly temperature in °c and rain fall in mm)
Wet months
Aridity limit
Water deficit
Water recharge
Water surplus
Water utilization

FIG–13

the state. This part comprises of the Hadoti Plateau which receives maximum amount of rainfall i.e., 25 to 27" annually. A isohyet line joining Banswara in the South, Kota-Jhalawar and Bundi in the SE and Bharatpur in the East, distinctively draws this part seasonally a different form the rest of the state. The annual budget of the precipitation accounted for about 65.7 cm against the 141.8 cm of total deficit amount clearly represents the moisture conditions in the soil and the extent of aridity limit (Fig.13-G and H).

The aridity limit and the water balance incidences may be calculated as in table 15.

The extent of aridity limit is well marked from June/July to September . At least, there are four months which are reducing the aridity limit, though, actually 3 months are marked in over the limit of aridity. The precipitation in July-August and September extending beyond the total losses of water i.e., 42 cm against 65 cm.

The water balance for the period well represents the recharge condition starts from June and utilised upto October thus, the moisture period extended from June to October i.e., 5 months, reveals that there is no need of watering the crops right upto 5 months (June to October) therefore, irrigation is needed particularly in the months of Dec., January and February while the Rabi crops are in the fields. The actual conditions of evaporation and precipitations are recorded in the table (Fig.13 H).

The reason for the longer period of moisture retaining in soil with higher rate of evaporation and transpiration is being black cotton soil. It adjoins the black soil regions of Madhya Pradesh and Gujarat states in the south and southeast respectively. Characteristically, black soil retains moisture for a longer period inspite of the higher rate of evaporation, they do not allow water to be let out. Therefore, the moisture is retained underneath the soil layer which helps in maturity of plants in absence of irrigation or rainfall. But, the area is broken out very severely during deficit months i.e., March to May, due to high sun and heating of the ground.

Ajmer district, part of Aravalli region having an altitude of 736 M, from sea level provides a case of 2 wet months in the northern apex of Aravallis, while in the southern part where it falls in the Udaipur district 3 wet months are experienced. The average of precipitation in this

TABLE-15

JAN	FEB	MAR	APR	MAY	JUN	JUL	AUG	SEP	OCT	NOV	DEC	
17.8	20.4	26.2	31.7	35.8	35.1	30.4	28.6	29.3	28.0	22.6	18.6	ct
0.7	0.8	0.9	1.1	1.5	1.4	1.0	0.9	1.0	1.0	0.8	0.7	Δtk
18.5	21.2	27.1	32.8	37.3	36.5	31.4	29.5	30.3	29.0	23.4	19.3	tk

From the Table I, it gives monthly aridity limit of

JAN	FEB	MAR	APR	MAY	JUN	JUL	AUG	SEP	OCT	NOV	DEC	
45	56	76	98	111	110	92	85	88	83	63	50	mmr

The period of day light is obtained from table-III

JAN	FEB	MAR	APR	MAY	JUN	JUL	AUG	SEP	OCT	NOV	DEC
-2	-2	+1	+2	+5	+8	+6	+2	+1	+1	+7	-1

So the precipitation is sufficient only when it is over

JAN	FEB	MAR	APR	MAY	JUN	JUL	AUG	SEP	OCT	NOV	DEC	
46	54	77	100	116	118	98	88	100	84	62	49	mmr

The actual precipitation

JAN	FEB	MAR	APR	MAY	JUN	JUL	AUG	SEP	OCT	NOV	DEC	
8.0	5.0	4.0	5.0	10.9	70.6	271.5	257.1	129.0	13.7	6.0	5.0	mmr

unit is about 22 inches or 52.7 cm with average temperature of 23°C. The monthly range of precipitation and the temperature conditions for the aridity limit are marked in Fig 13. CD and EF.

CONDITIONS OF HUMIDITY

Balancing of Hydrologic Budget

To determine hydrologic budget is a bit different than the aridity limit and wet month limitations. Actually it is a technique to determine humidity variations from month to month and to fix limits of humidity, e.g., semi-humidity, aridity and semi-aridity with the help of set method. It is also to formulate reference numbers for humidity conditions to establish humid zones. Because, the hydrologic balance of a region, to the extent of aridity or humidity limits, helps in determining the feasibility of irrigation and water requirement of plants. It sets the condition of optimum water balance in soil as moisture and probable range of its utilization. The present scheme of demarcating the boundaries of humidity/aridity zones was tried and interpreted by Penk (1910) who further extended the work of Thornthwaite and Mather (1957), for calculating factors of the hydrologic balance with minor amendments. Though, Penk's misconception of, "Precipitation equals evaporation" (P = E),was corrected later on.

For the present study, Wolfgang Schmiedacken's model for calculating water in the soil made available to plants, has been applied here. Water at the root of plants may by calculated from the following formula:

$$St^n = \frac{St_n - 1}{\dfrac{(PEt - P)\ n}{Wc}}$$

Where :

St_n or Stn -1	=	Volume of water in the soil at the time (month)
Wc	=	Water holding capacity in root zone of soil
P	=	Precipitation
PEt	=	Potential regional evapotranspiration

The water balancing budget may be calculated from the following set of determinents, given in the table:-

TABLE-16

KOTA DISTRICT : Water Balancing Table

	JAN	FEB	MAR	APR	MAY	JUN	JUL	AUG	SEP	OCT	NOV	DEC	
P	8.0	5.0	4.0	5.0	10.9	70.6	271.5	257.0	129.0	13.7	6.0	5.0	mm
PEt	70	76	107	137	164	163	148	137	135	122	90	69	mm
P-PEt	-62.4	-71	-103.1	-132	-153.1	-92.4	+123.5	+120	-6	-108.3	-84	-64	mm
St	4.9	3.7	2.9	4.0	1.9	95.5	217.6	201.3	200	10.1	4.0	3.2	
P/PEt*	0.1	0.06	0.03	0.03	0.06	0.4	1.8	1.8	0.9	0.1	0.06	0.07	
P-AL	-36	-47	-73	-95	-106	-37.4	+173	+169	+29	-70.3	-56	-44	mm

* Humidity Reference Number is 0219; i.e., 9 Arid Months, 1 Semi-Arid, and 2 Semi-humid Months. (Pl-see scale).

TABLE 17

JODHPUR : Water Balancing Table (Arid Zone)

	JAN	FEB	MAR	APR	MAY	JUN	JUL	AUG	SEP	OCT	NOV	DEC	
P	3.6	6.0	2.7	3.2	10.0	35.5	99.0	122.0	6.0	8.0	2.7	2.7	M
PEt	80	91	115	139	159	160	148	137	138	127	104	86	M
P-PEt	-76.6	-85.0	+12.3	-135.8	-14.9	124.5	-49.0	-15.0	-78.0	-119	-101.3	-83.3	
St	2.1	3.6	1.7	2.2	22.8	32.2	198	74.8	70.7	5.5	1.6	1.5	
P/PEt	.14	.10	.02	.02	.06	.02	.6	.8	.4	.06	.02	.03	

Scale of humidity/aridity Measurement of Reference numbers at Jodhpur.

>2 humid	Nil
1-2 semi humid	Nil
0.5 - 1.0 Semi- Arid	2 months - July, August
<0.5 Arid	10 months

Maximum amount of water i.e., 748 MM per sq. c. meter is available to plants in the month of August.

The above water balancing table may be explained as under:

Line 1 Gives the mean monthly precipitation -P.
Line 2 Shows the mean monthly potential regional evapo-transpiration PET = $(E_{mx}-E_{min-2})$.
Line 3 Gives the climatic hydrologic budget i.e., the difference between the precipitation and potential regional evapo-transpiration.
Line 4 Calculates the water in the soil and still available for use by plants. The equation is explained earlier.
Line 5 Represents the Ratio of precipitation to potential regional evapo-transpiration for reference numbers.

While calculating the limits of aridity, the limit of hydrological balance for humidity zone forms an essential part of the present study. It is through the rate of potential evapo-transpiration and the amount of precipitation, the actual conditions of the moisture in the soil is ascertained which is necessary for the growth of plants. Thus, by determining the range of humidity or aridity one may deduce the extent of water amount in soil and the requirement for plants and can draw Isohydrostatic lines with the help of "Reference Numbers". Though, it needed a separate study.

The reference number for the area may be calculated from the line 5, as has been explained in the case of Jodhpur district in the preceding pages, while determining the humidity/aridity zones. Actually, it is the ratio of the precipitation to potential evaporation. This may be expressed as 0021, means, two semi-arid months and ten arid months. No humid and semi-humid months are recorded at Jodhpur.

Further, it may be explained by taking an example from Kota district : (see table).

There are two semi-humid months (July-August), one Semi-arid (Sept.) and nine arid months. All these months may be arranged in the reference number of 0219. The above table also explains the water conditions in soil and its changing ability. Line 4, reveals the water at the root zone of plants, i.e., maximum 217.6 MM per sq. c. meter of surface water is available to plants in the month of July. Similarly the change in soil, water also may be calculated, which occurs very frequently due to the changes in the atmosphere, i.e., temperature and rainfall etc. Water in the soil creates the necessary conditions for the surface changes i.e., Run-off, evaporation, and percolation process etc. The changes in soil-water, is actually the difference of precipitation and the aridity limit which is already set in for the area. Because the existing aridity limit; the extent to which water is required and the amount of precipitation actually received makes the soil moisture or dry for the period. Soils get wetness or moisturity over and above the aridity limit, unless the precipitation exceeds the limit, no change is brought about.

The line 6 reveals that July, August and September are the months which receive more precipitation than the aridity limit causing changes in the soil water upto +173, +169 and +29 MM respectively. This water may be termed as surplus water for the period, available for plants. Whereas in rest of the period, aridity over-rides the precipitations, all are recorded in minus values, makes the entire period as dry and no change in soil water is expected. Though the water budget graph extents the water utilization limits, but does not reject the hypothesis.

If crop associations are drawn in accordance with the hydrologic budget alongwith aridity limit, one may deduce the greater degree of dependency and relationship. The area which are well known for their Mono-crops as Bajara or Jowar are falling within the fold of long periods of dryness with single wet month. In contrast to it where more than five or six crop associations are located found with longer period of wetness at least 3 months, alongwith sufficient moisture in the soil lasting for five months. In such area variety of crops are grown or cultivated, but with the support of irrigation net work. Thus, the aridity and humidity limits set the conditions of water balance, necessary for plants and help in locating the area of water scarcity for which irrigation becomes an essential part of it.

CONCLUSION

Out of the above study one may reach to the conclusion that the aridity and humidity are the essential conditions for the plant's generic process. The precipitation and evapo-transpiration are the controlling factors of aridity limit or humidity conditions, which in turn affect plant's growth and generate conditions of agricultural development. Thus, soils also become integral part of the whole system in which hydrological conditions deter the regional characteristics. Even, then, it is the amount of precipitation and the rate of evapo-transpiration, which determines the regional water budget or hydrological balance, on which the excess of humidity or aridity depends, necessitate water requirement of plants. This factor involves the water regulation and if it is inadequate, then it is met by irrigation system.

Jodhpur and Kota are the two extreme cases in Rajasthan which provide the examples of the excessive aridity and mildness respectively, brings on the line that besides, continental efforts with regard to temperature and rainfall studies, local variations also contribute much towards the growth and development of plants. Thus, if other areas of the Indian States are extracted on the similar lines, they may be of great help in demarcating the Agro-climatic regions.

TABLE-I

Monthly aridity limit in mm of rainfall with corrected mean monthly temperatures (t_k) ignoring the length of daylight

t_k	0.0	0.1	0.2	0.3	0.4	0.5	0.6	0.7	0.8	0.9
0	0	0	0	1	1	1	1	1	2	2
1	2	2	2	3	3	3	3	3	4	4
2	4	4	4	5	5	5	5	5	6	6
3	6	6	6	7	7	7	7	7	8	8
4	8	8	8	9	9	9	9	10	10	10
5	11	11	11	11	12	12	12	12	13	13
6	13	13	14	14	14	14	15	15	15	15
7	16	16	16	16	17	17	17	17	18	18
8	18	18	19	19	19	19	20	20	20	20
9	21	21	21	21	22	22	22	22	23	23
10	23	23	24	24	24	24	25	25	25	25
11	26	26	26	26	27	27	27	27	28	28
12	28	29	29	29	29	30	30	30	30	31
13	31	31	32	32	32	32	33	33	33	34
14	34	34	34	35	35	35	36	36	36	37
15	37	37	38	38	38	39	39	39	40	40
16	40	40	41	41	41	42	42	42	43	43
17	43	44	44	44	45	45	45	46	46	46
18	46	47	47	47	48	48	48	49	49	49
19	49	50	50	50	51	51	51	52	52	52
20	52	53	53	53	54	54	54	55	55	55
21	55	56	56	56	57	57	57	58	58	58
22	59	59	59	60	60	60	60	61	61	61
23	62	62	62	63	63	63	64	64	64	65
24	65	65	66	66	66	67	67	67	68	68
25	68	69	69	69	70	70	70	71	71	71
26	72	72	73	73	73	74	74	74	75	75
27	76	76	76	77	77	77	78	78	79	79
28	79	80	80	80	81	81	82	82	82	83
29	83	83	84	84	84	85	85	86	86	86
30	87	87	88	88	88	89	89	89	90	90
31	90	91	91	92	92	92	93	93	94	94
32	94	95	95	96	96	97	97	98	98	98
33	99	99	100	100	100	101	101	102	102	102
34	103	103	104	104	104	105	105	106	106	106
35	107	107	108	108	108	109	109	110	110	110
36	111	111	112	112	112	113	113	114	114	114
37	115	115	116	116	116	117	117	118	118	118
38	119	119	120	120	120	121	121	122	122	122
39	123	123	124	124	125	125	126	126	127	127
40	128	128	129	129	130	130	131	131	132	132
41	133	133	134	134	135	135	136	136	137	137

TABLE-II

Determining the additional temperature in °C to allow for the altitude when calculating the aridity limit according to Schreiber

Mean monthly	Altitude of meteorological station in metres									
temp.°C	500	1000	1500	2000	2500	3000	3500	4000	4500	5000
0	0.9	1.8	2.7	3.6	4.5	5.5	6.6	7.6	8.7	9.7
2	0.9	1.9	2.8	3.8	4.8	5.8	7.0	8.0	9.2	10.2
4	1.0	2.0	3.0	4.0	5.1	6.2	7.4	8.5	9.7	10.8
6	1.0	2.0	3.1	4.2	5.4	6.6	7.8	9.0	10.2	11.4
8	1.1	2.2	3.3	4.4	5.7	6.9	8.2	9.4	10.7	12.0
10	1.1	2.2	3.4	4.7	6.0	7.2	8.6	9.9	11.2	12.6
12	1.2	2.3	3.6	4.9	6.3	7.6	9.1	10.5	11.9	13.4
14	1.2	2.4	3.8	5.2	6.6	8.2	9.6	11.1	12.7	14.3
16	1.3	2.6	4.0	5.5	7.0	8.6	10.1	11.8	13.5	15.2
18	1.4	2.8	4.2	5.8	7.3	9.0	10.6	12.5	14.3	
20	1.4	3.0	4.5	6.1	7.7	9.5	11.2	13.2		
22	1.5	3.2	4.7	6.4	8.0	10.5	12.0			
24	1.6	3.4	5.0	6.7	8.4	10.5				
26	1.7	3.6	5.3	7.0	8.9					
28	1.8	3.8	5.6	7.3						
30	1.9	3.9	5.9							
32	2.0	4.1								
34	2.1									

TABLE-III

Percentage additions and subtractions to allow for length of daylight(S)

Latitude	J	F	M	A	M	J	J	A	S	O	N	D
0 degrees	1	1	1	1	1	1	1	1	1	1	1	1
10	-3	-2	1	3	5	5	5	4	2	-1	-3	-4
20	-8	-4	0	5	10	11	10	7	2	-3	-7	-9
30	-13	-8	0	8	14	18	16	10	3	-4	-12	-15
40	-19	-11	-1	11	20	25	23	15	4	-7	-17	-22
50	-28	-15	-1	15	28	35	33	21	5	-10	-25	-32
60	-44	-23	-2	22	43	55	50	30	8	-16	-38	-50
70	-89	-41	-3	38	87	100	100	52	13	-28	-69	-100
80	-100	-70	-5	80	100	100	100	90	25	-60	-100	-100
90	-100	-100	-60	100	100	100	100	100	80	-100	-100	-100

NINE

Water Harnessing Problems as a Basis for Development

ORGANISING IRRIGATION PROJECTS

To meet the growing challenges of rapid strides of population growth and supplying of agricultural products is one of the most important task of the future in most of the developing countries like India. It may be assumed for the present decade, at least that rain-fed agriculture must be given greater emphasis even then, in addition to it, however irrigation will have to ensure greater role particularly in the areas of flooded water just as to prevent devastation on one way and to utilise water for crops on the other. This challenge can be met out only through the construction of irrigation projects.

But, heavy investment is accurred not only on construction of new projects, but also on maintaining old ones, and thereby it becomes a national subject. This fact invites proper planning and layout of the project which is to be commissioned in any area.

Before going to install a project, one has to look into the material needs based on certain inputs of organisational structure and converted into out-put by transform processes. Input systems comprises of two aspects i.e., people and resources whereas, output system embraces with categories of goods and services. The former involves the labour and their services in various forms and the material resources required for a project. These together form action units employed with a technology to produce power and utilise water for irrigation.

The latter, output systems, consists of agricultural products for human and animal food, and raw materials for processing.

While doing so, every irrigation project is faced with a particular situation. This situation may be understood as the environment of the irrigation system, embraced with socio-cultural systems of a village community by its inter-alia specific social structure, value systems of an individual and social structure; tibes and social groups, codes of behaviour, distribution of status and roles. The social and cultural milieu of an irrigation organisation influence the whole structure. In the environment of an irrigation system the ecological set-up takes an important consideration rests on the natural factors like rainfall temperature, nature of soil, topography and availability of water etc.

Irrigation as a basis for development may be regarded as a factor of production to improve the economic conditions hence, it has two considerations :

- Water can be regarded as production factor, it acts as a means of increasing the yield per unit of area.

- Irrigation increase the possibility of food production.

The present study is to focus on irrigation problems while stressing on the 'auxiliary inputs'; water and crop yield relationships, some other considerations have to be kept in mind, such as

- reliability and required accuracy.
- cost and time requirement.
- flexibility.
- intrutive power.
- calibration etc.

Because, there is no reliable mechanical approach, where uncertainty in selection of inputs, and demarcation of fixed timing can not be established for any project. In other words it may be taken that the elements of project planning depends upon the following factors : (Applied Geog. Vol. 23, pp 94-102)

- Agricultural programme.
- Water requirements.
- Ratio of use of the irrigation water.
- Farm irrigation demand and irrigation water.

Besides all that, optimisation of irrigation is a crucial factor which needs so many considerations while establishing any project. In optimisation of irrigation; the initial phase of planning starts with the supply of water that is 'supply schedule' which means determination of the necessary water supply and its duration, including the frequency and the rotation of supply to a field.

With the objectives of water supply, a structure has to be erected in which water may be stored to regulate timely discharge. This involves a huge task and an amount, which may be realised from some of the cases.

The cost estimation on big projects, as well as on medium projects may well be visualised from some of the projects schemes, either have been completed or under completion in the state. The well known and the biggest irrigation project in the state is Indira Gandhi Canal, which was launched in 1948, but the actual work could be commissioned in 1950, with the initial outlay of 66 crores, which rose to 405 crores at the end of the financial year 1984. It is estimated that the complete project would be required about 1085 crores of rupees. It has been constructed in two phases; one as the 'Rajasthan Feeder Canal' in Punjab state and second as Rajasthan Canal, now Indira Gandhi Canal, with its 2950 km. distributary systems in the state itself. The length of the main canal, right from the Harike barrage in Punjab to Jaisalmer (upto Gadra Road) is about 9,425 Kms. with this size, one may think of gigantic project lying in the desert areas of the state and its utility on completion.

Similar is the case of another giant venture in the state witnessed by the construction of three consecutive Dams on the Chambal river known as 'Chambal Valley Project Schemes'. This scheme too, is one of the national level schemes jointly sponsored by M.P. and Rajasthan in 1943 at Kota itself, but later on, expanded to three Hydro-projects in 1953-54. The total outlay of this scheme was estimated as about 100 crores, but later on splited into three respective projects. Kota Barrage itself has about 2350 Kms. long distributory systems. But these projects in turn posing some problems of irrigating the fields.

The another developing pace of irrigation projects is provided by the gigantic layout of Mahi Bajaj Sagar Project, Bhakhra Dam, and many other small project schemes, which have been mentioned in the preceding pages. Though these projects may be considered on a state level, but with a wider perspectives of canal irrigation with distributory systems involving crores of rupees. In the perspective planning how far these schemes are beneficial and world serve the right cause of the laid down objectives, yet to be seen in time to come.

Planning perspective should match with the existing conditions and the local requirements of the cultivate plants in fulfilling the needs of the people. It may well be borne out that certain crop/plants have different needs of growth and development, as it is clear from some of the case studies of the cultivated crops in the foregoing pages. Even then, observations show that some crops having different root-systems flourishes differently in their eco-system. The requirement of water, soil characteristics and moisture conditions, temperature and water regime of the soil etc. may be considered as forming a complex combination of elements known as eco-system. Examples have been extracted from some of the areas revealing the crop-association with their environment.

The general requirement of temperature for various crops is as follows:

Spring wheat	0-4°C	Winter Wheat	10°C
Oats	6°C	Millets	
Field Corn	12-14°C	(Jowar/Bajra)	18°C
Sweet Corn	10°C	Maize	18°C
Potatoes	7°C	Barley	12°C
Peas	4°C	Sugarcane	17-24°C
Cotton	17-18°C	Rice	20°C

If excess temperature is occurred at a place then the difference of required and the excess temperature may be regarded as causing much need of moisture in the soil. If temperature in particular month exceeds the actual requirement then there will be loss of moisture from soil, and soil will be moistureless (deficit). If the temperature is below the requirement then it may be regarded as there is no need of additional moisture in the soil, which is already exist (sufficient) is called the 'moisture potential range of soil'. Thus the amount of precipitation in a calendar year should have been on or above the mean of the station (area) for the normal growth of the plants/crops.

CASE STUDIES

In this context some particular cases of crops with

regards their moisture conditions, root systems and vegetive periods are highlighted as under:

Among cultivated crops, lucerne has the longest root system i.e., about 2 metres and able to suck water from the 'A' horizon of the soil layer. While other crops get water only from the Litter or fermented layer of 'Ao' zone, and therefore, find it difficult to survive in absence of moisture. In high temperature water evaporates very quickly from the litter and fermented or even from Humified layer of Ao zone. Thus, irrigation becomes necessary for almost all the crops.

Lucerne is a green fodder for animal, cultivated extensively round the year in all parts of the state, where water is available abundantly. It is a sensitive crop to water, and flourishes continuously in all the seasons for 3 years after seedling once. But quality depends upon the supplying of water. It needs water very frequently depending upon the soil quality at least once in a week in summer and fortnightly in winter. It is observed that if water is given in the early hours of the morning i.e., at about 4 a.m. it thrives well and growth is faster. If water is given in the day hours it does not yield well and plnat is near to the wilting stage. It is because , that the day irrigation is not helpful in metabolism process of the plant. Secondly most of the water in evaporated and thirdly, the severity of winter cold in the early morning hours absorbs soil temperature and hampers growth badly which is regained through water.

Similar is the case of wheat and barley crops, which needs water in the early morning hours, so that photosynthesis process may be accomplished in the day light sun for the better and rapid generic process and development. If water is supplied in the night or late evening after the sun set, the process is not accomplished due to lack of sun light and plant banishes to flourish. In addition, such plants don't' preserve sufficient heat in soil to survive against the severity of cold-nights in winter. In night irrigation the losses of soil heat are greater and thus, plants find it difficult to survive or yield luxuriantly in absence of metabolism; a process of dissolution of plants' food substances in presence of sun-light.

It is evidently clear that farmers in village sites may still be seen today on fields in the early hours of the morning pulling water to irrigate the fields. It is a customary practice. Thus, these facts prove our assumption that night irrigation for certain crops is helpful in yielding more and better production.

But on the contrary, an example of irrigating Gram at night proved to be a futile result is from the Rajasthan Canal area of Sriganganagar district, reported by Saran, a research scholar. He observes that a farmer wanted to replace scattered gram plants by onion in his small patch of field, so he provided the water at 12.30 a.m. In the following morning, it was noticed that the plants were badly shrinked with cold and scorched severely. This fact proved that gram plants need not to supply water at night. Because, due to night watering the soil lost its preserved heat and supplied water became cool in the morning hours due to further lose of heat. This strengthens the assumption that night irrigation for certain plants is not favourable.

But for Rice and Sugarcane, things are different as both the crops need maximum amount of water and at times field should be filled up with water, so giving water at night carries no meaning. The large amount of water preserves sufficient heat within soil regime to prepare food for plant protection. Thus, it is a factor of physical requirement; climate, soil etc., and the biological needs like root systems, metabolism etc. for different crops. The conditions provided by the ground water are altogether different one.

The problem of wide spread salinity in the Wn. parts of Rajasthan, is perhaps less visualised, so far its consequences are concerned. But some efforts are being made by CAZRI of Jodhpur in bringing out those areas which may yield better results against salinity on irrigation along the old courses of the rivers.

Photogeomorphological studies have revealed the distribution of saline lands in Rajasthan desert along the old courses of the drainage channels, especially along their lower reaches or at their confluences. These channels have played significant role in the evaluation of agro-irrigation practices in the lower parts of the old courses of the major river basins the saline problems have been cropped up and causing concern to all of us.

Studies made by some scientists at CAZRI, Jodhpur with special reference to Bimal Ghose, S.Singh, and Amal Kar's work are worth-mentioning in this regard. These scientists in their efforts have highlighted some aspects of salinity hazards in Wn-Rajasthan, and have tried to trace out all those areas which have been affected due to the agro-irrigation practices along the river basins more particularly at the Luni river.

They are of the opinion that the construction of canals across the old drainage channels and irrigation of these areas by canals have substantially contributed in the development of salinity. These patches may be seen around Manhawa, Mori, Sutran and Kharabera, Biotan in the Swn. Jodhpur where old canal crosses an old bed of the Luni.

These scientists have provided a map showing the saline patches along the old courses of the main channels in Jodhpur and Jaisalmer districts, throw sufficient light on the problem (Figs.14-15).

The Luni Basin and Irrigation

The irrigation and misuse of water are the two apex of the water management system. In desert areas of Rajasthan where water is the more precious and scarce commodity the judicious use of water is the supreme concern. In the areas of meagre and unpredictable flow of surface, water generally goes as waste, human efforts are diverted to use it through suitable construction of embankments. It needs no further explanation that many of the large irrigation reservoirs in the fragile eco-system of the desert have been erected without much consideration vironmental perspectives.

The Luni basin provides most of the unwarnted and unwaned such construction which lead to unprecedented problems of water management, for which a survey was conducted by a team of scientists in 1981-82. For the purpose the upper Luni basin with a plethora of channels, presents one of the most integrated drainage net work in the Wn. part of Rajasthan. It has an area of 13440 sq.Km. covering parts of the Aravalli mountains in the east and the desert plain in the west.

After going through the surrounding physiography, drainage pattern, soil composition , and many other environmental parameters with the stream characters, limiting by increased aridity and sandy moisture of the deposits, some facts have been highlighted by these scientists in the following way so far the better water management is concerned.

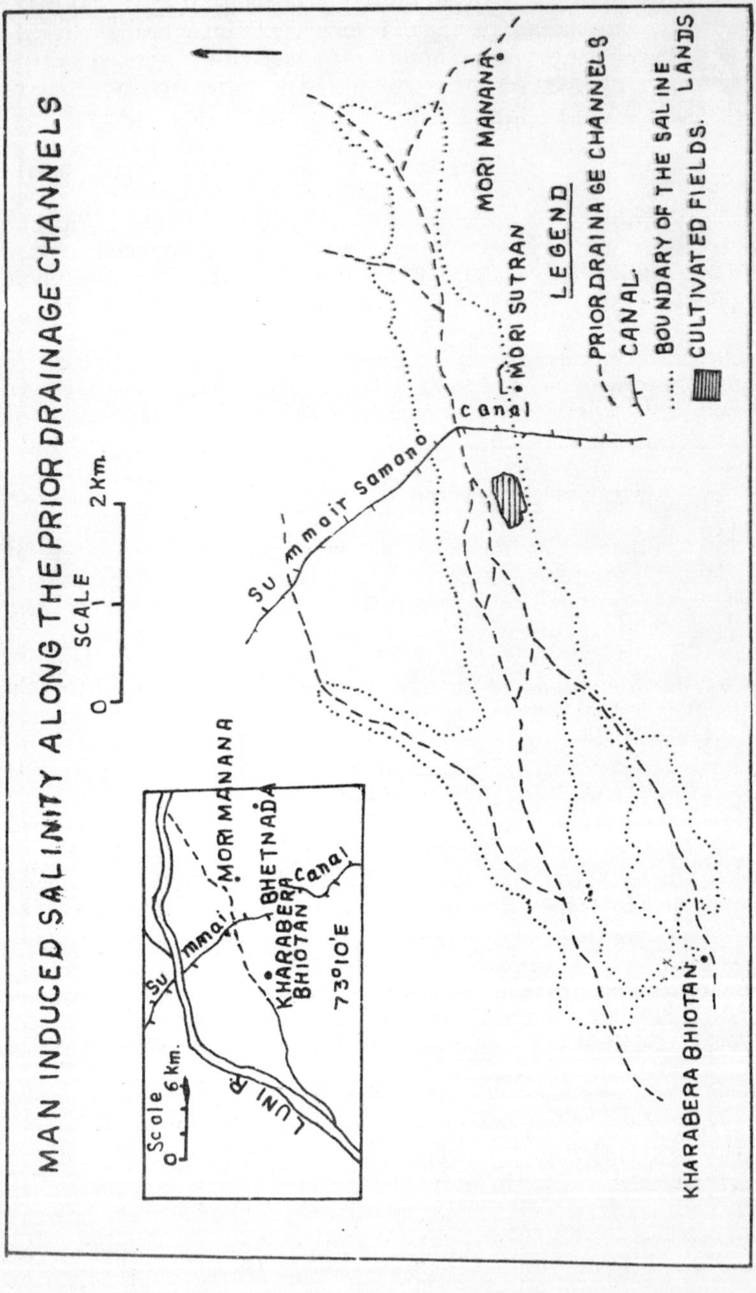
MAN INDUCED SALINITY ALONG THE PRIOR DRAINAGE CHANNELS
SCALE
0
2 Km.
MORI MANANA
MORI SUTRAN
canal
Summair Samono
KHARABERA BHIOTAN
LEGEND
PRIOR DRAINAGE CHANNELS
CANAL.
BOUNDARY OF THE SALINE LANDS
CULTIVATED FIELDS.
Scale
0
6 Km.
LUNI R.
Su
mmai
MORI MANANA
BHETNADA
KHARABERA BHIOTAN
Canal
73°10'E
Bimal Ghose, S. Singh and Amal Kar

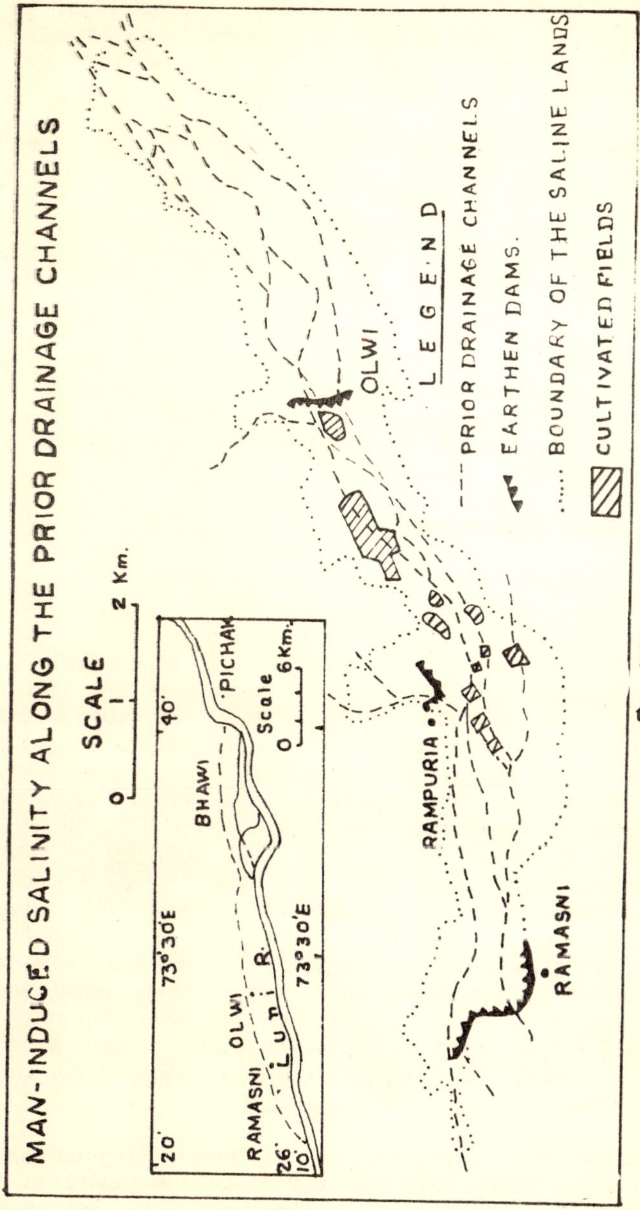

Bimal Ghose, S. Singh and Amal Kar

Fig. 15

(a) Profuse irrigation and lake of proper management of the reservoir water at Pichak and Sardar Samand on the Luni, have created salinity problems around Bhawi and Sardar Samand Farm area. Salinity has come up at an alarming rate in the latter case.

(b) The Kharda reservoir on a granitic terrain to the north west of Pali is another area accelerated salinity development. In between Kharda and Bandai feeding water from Sardar Samand lake has created more serious problems of salinity development because of reservoir itself.

(c) A different type of problem is coming up along the Bandi downstream of Pali, because of continuous use of the river bed by the local dyeing and printing work and regular dumping of the industrial wastes into the river, have greatly increased the toxicity level of the water.

Socio-Economic Constraints

But at the same time, some cases are of worth-mentioning in the light of economic considerations. The well planned field parcels and irrigation channels may provide the better stimuli to the crops. The example of this type may be cited from Sriganganagar district itself, which has a planned irrigation system in Chaks (Fig.16). Here the water is regulated as per the requirements of the crops in time. But, again as it is mentioned earlier that the Fery system is strictly maintained in otherwise, dead bodies may be seen floating in the canal water at times.

The cropping pattern; Rabi, and Kharif, is set with the crops of wheat, cotton, sugarcane, rice and millets. Due to irrigation rice is seen with changing pace with the time. For that Sriganganagar provides the case of better management of water distributories and the cropping relationship to the time, with some experiment of newly inducted Hyv seeds. All this could be achieved due to ample water to desert soil which responded favourably in all adverse conditions of temperature and precipitation.

A similar but a distinct case of water harnessing the crops, may be provided from a village Jod-Kandra in Pali tehsil of Pali district. This village has a typical case of cropping in Kharif and Rabi seasons. There is no cultivation in Kharif as the whole area is filled-up with rain water and

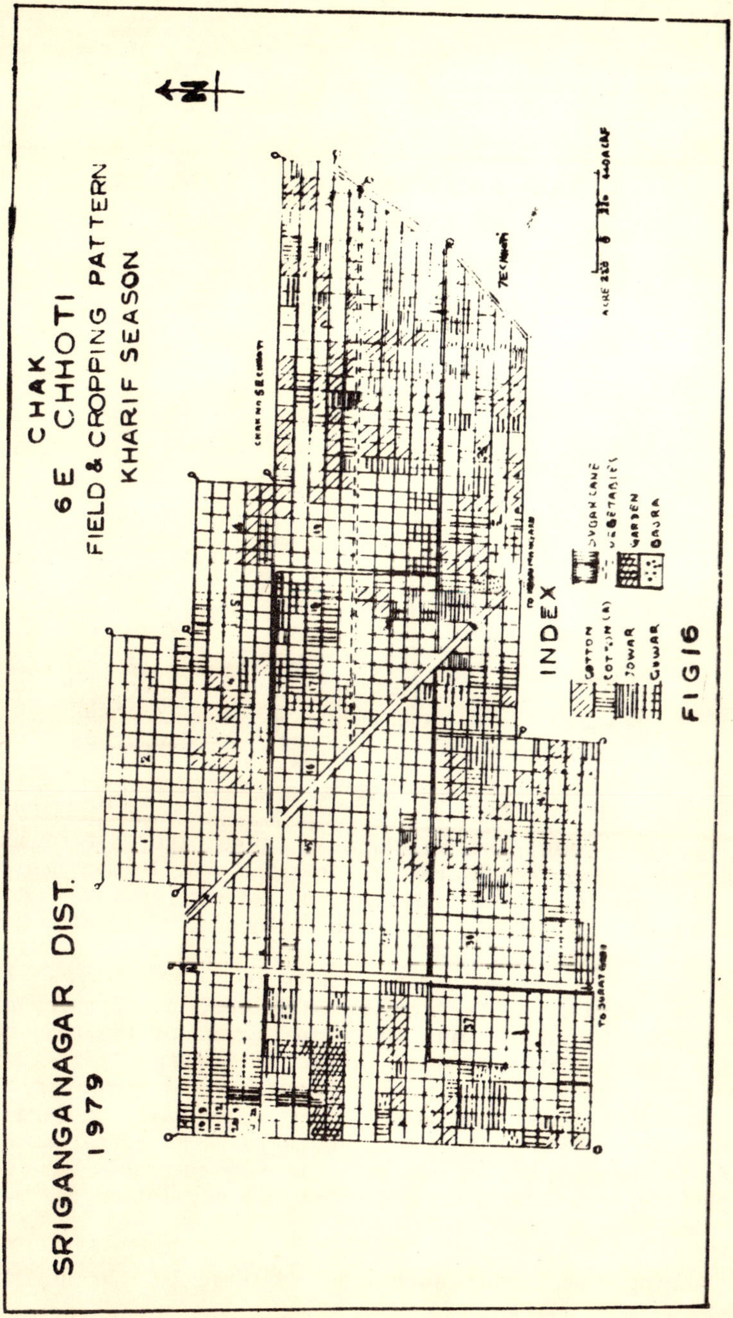
SRIGANGANAGAR DIST.
1979
CHAK
6E CHHOTI
FIELD & CROPPING PATTERN
KHARIF SEASON
INDEX
COTTON
JOWAR
GUWAR
SUGARCANE
VEGETABLES
GARDEN
BAJRA

FIG16

turns to be a swampy. But, in Rabi all the irrigated crops are grown luxuriantly and land is used intensively (Fig.17). Thus, it provides an altogether different case of water and crop relationship. No other area in the vicinity of it, exercise this type of practice, as the entire region falls in driness through out the year.

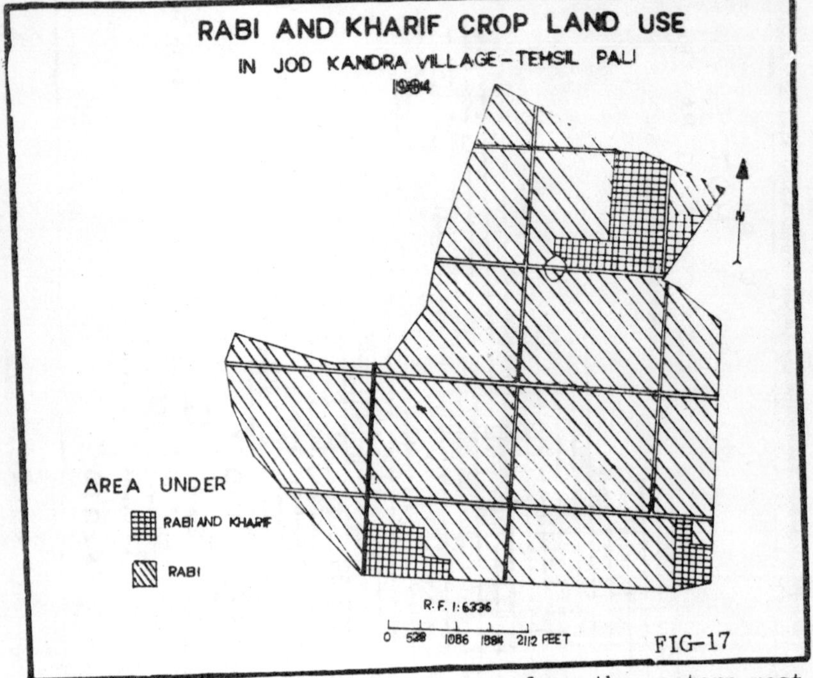

FIG-17

Quite in contrast to these examples, the eastern most districts; Alwar and Bharatpur, provides a different case of water harnessing. This area is well known for variety of crops especially wheat, sugarcane, oilseeds, tills etc. etc. It is a fertile track of Trans-Yamuna plain with abundance of moisture in the soil. Water table too is not below at the surface, with these favourable conditions, the farmers are in favour to adopt mustard crop instead of wheat as the water requirement of mustard is less than wheat. There is no need of frequent irrigation to mustard and moreover, it fetches more money than wheat. Thus economically mustard is more viable. With these the cropping pattern in the eastern part is changing day by day with the easy adaptable cum less expensive crops. No doubt other regions may follow the suit in some year to come.

During the survey author noted some of the points

regarding the crops in relation to their environment and observed that most of the crops are unfoundedly being cultivated just as they have been cultivated by their ancestors, and hence, there has been no change in the cropping pattern too.

It is observed that cotton is extensively cultivated in Bhilwara districts and in some neighbouring areas despite the fact that these areas practically do not provide favourable conditions as required for it. Cotton cultivation in this area is simply a tradition and customary practice to obtain a bit case amount of it as some cotton processing and spinning mills are located at Bhilwara and its vicinity. Though the quality of cotton is poor and staple is short, meant for rough clothing only.

On the other side, this area is most suitably favour the cultivation of sugarcane. The old establishment of Bhopal Sagar-Sugar Mill by the then Mewar Maharaja Shri Bhopal Singh Ji at Bhopal Sagar near Chittorgarh, is well be recounted for his farsightness looking to the hinterland as sugarcane growing area.

The construction of 'Mezan Dam' at about 4 Km. away from Bhilwara town has additionally attracted the cane-grower along with cotton. But the inclination seems to be more towards cotton cultivation rather than sugarcane as it is not very much lucrative crop. Being bulky it incurs more transport cost. Therefore, inspite of favourable conditions, farmers do not cultivate it. But, in turn if it is cultivated and sugarcane crushing mills are established near by area, this region may be an 'another Meerut of Rajasthan.

The completion of 'Mezan Dam' has added many other attractions to the farmers to grow variety of crops in addition to the existing ones. In the vicinity of the Mezan Dam command area, channels are favoured with topography and nourished by climate, may help in cultivation of economic crops.

From the above facts, though it is very difficult to relate the exact need of water for different crops due to changing environmental conditions and increasing complexities of eco-system; even then, the study may be helpful in delineating some of the areas of crop-water relationships to stress upon the better harnessing conditions, as it is shown in the table below.

TABLE-18

SOIL - WATER AND CROP ASSOCIATIONS

Soil Types	Sub Types	Texture of soil	Composition	Defficient in mineral	Moisture Retaining capacity	Vegatative period of crops	Main Crops	Water requirements	Root Systems	Main Areas of cultivation
Desert	Grey & Brown or Ferruginous	Ca CaO_3 (Kankarbed)	ca, mg, and carbonates	Humas Ni	Highly Permeable (Less)	3 to 5 months	Jowar, Bajara, Maize, Pulses & Fodder	V.High	Very short or V.long for Trus.	All the districts in the Desert Fold(11)
Red & yellow	Loamy clay	lighter	lime, K,P iron-oxide	carbonates Humas	good porous (moderate)	4 to 6 for some 8 - 10 months	Wheat Maize Barley Millets, Gram, Pulses Sugarcane, Fodder	High (mostly irrigated)	Medium to low	Udaipur Bhilwara, Ajmer
Black	Red & Black (Mixed)	Medium Loamy clay	Nitrogen iron oxide P,K Ca	Humas	V.High	4 - 6 months even 10	Wheat, Barley, Cotton, Sugarcane, Pulses Lucerne Fodder	Medium to high	Medium to long	Chittor, Dungarpur, Banswara, Kota, Jhalawar, Bundi.
Alluvium	Alluvial Eluvial & illuvial	Clayey Loamy Slit	Humas, Na, Ca, mg, etc.	lime Ni, P	Medium to High	6 - 10 months	Wheat, Barley Sugarcane, Pulses Mustard, Till, Rice etc.	Low to Medium	Medium to long	Alwar, Bhartpur, Parts of S.Madhopur, Jaipur, and Tonk

TEN

Conclusion

Some recurrences may be drawn from the preceding pages that how plants are susceptive in their response to water requirement depending upon the quality of the soils. i.e., soil formation, structure and composition. The highlighted areas of the water scarcity on the basis of the humiding conditions and the limits of Aridity in the state, are reflective of the plant growth in meeting with adequate amount of water in a particular cropping period. If the study concluded on this basis then one may come across the following assumptions:

1. The soils of the South-Eastern Rajasthan belong to the Block Level group, with clayey particles makes the harden surface with higher water retaining capacity. On the other hand, the Aridity limit observed for the area, speaks for the higher precipitation rate at all the times at least 3 to 4 months surplus water, enable plants to grow frequently but those with stronger root systems, penetrating deep in to the soils.

   Though, the conditions are hardened in summer when the sun strikes hard, enabling plants to sustain upon the water available at the below surface level. Thus, the crops of strong and long root systems may thrive upon these conditions. While on irrigation, these soils yield favourably and produce a good result. But the Hadoti platean again harden the conditions.

2. In the Eastern parts, the Trans-Yamuna enclave form the most fertile plain or Alluvium cover in the parts of Alwar, Bharatpur, Jaipur, S.Madhopur and Tonk districts. But, the medium aridity limit, enabling water to the plants only for about 2 to 3 months, as surplus while in other months except summer mild moisture in the soil,

harness crops like weakly root system. Summer make the conditions more harden for the crops because of increasing aridity in the region.

3. The Western parts facing acute water shortage, with lighter particles of sand and meagre amount of rain limiting aridity for above 11 months. This makes hard to sustain crops upon such conditions, except in low lying areas of River basins. Evaporation rate is higher at all the times.

Crops of Short & Long Vegetative Periods

It is unlikely to reveal that most of the crops are grown in accordance with the aridity limits, that means those crops which have longer vegetative period do not grow in the areas of higher aridity limits and vice-versa. It is coincided that crops which have short vegetative period, take less time to mature, hence, need less amount of water. These crops thrive well within a short stipulated time when the moisture is sufficiently available. On the contrary, crops which take 6 to 10 months in their maturity, can be grown only in the areas of having shorter period of arid conditions, that means, much moisture round about 6 to 8 months with higher Humidity or surplus water for about 3 months.

These facts can very well be related with the crops of longer vegetative periods like sugarcane, jute flax, wheat, cotton etc., while those with shorter vegetative period jowar, bajara, maize, oilseeds, mustards vegetables etc., with the exception of rice, and lucerne, which takes shorter period with greater need of water. Of course the duration of vegetative periods depend upon plant's generic process but harnessed by soil characteristics enabling plant's early growth and development.

Therefore, it would be interesting to draw a line of common factors, such as Aridity limit, humidity conditions, soil characteristics and the vegetative periods of crops, to delineate the regions of water requirements and the crops in association with it.

Loses of water is greater in Sriganganagar, due to higher Trans-evaporation rate. But plants are heavily and sufficiently nourished with adequate amount of water, hence no crop damage is reported on account of water. But, in south-eastern part of the state, conditions are different as regards the water loses due to the faulty methods of irrigation as well as ill fated draining systems. Water is extensively breached-out from the canal viz; Chambal Canal and the area commanded by it, is facing water logging problems all along the left and right flanks. Though the loses of water due to evaporation is lesser, as being black-cotton region as Black soil retain moisture for a longer period.

In the Central parts of the Aravalli region, facing many fold problems of defective draining system as well as faulty methods of keeping water channels along the parcels. Though, there is no immediate remedy for it unless the reconsolidation of holdings is carried on. Most of the holdings are either fragmented or pieces hence, found with less than 1 ha. in size, find it difficult to get water in a mechanised way, particularly in the areas of undulating terrain like Udaipur, Bhilwara, Ajmer, Banswara, Dungarpur districts.

In contrast to these areas, Wn. Rajasthan, faces altogether different problems of watering the crops. This region is desertic in nature, hence, devoid of water and vegetation. Sandy soils posing another problems, though not infertile, but very thirsty and susceptive, needs different type of treatment while irrigating crops. Soils are unable to retain moisture for longer period. Hence, only crops of shorter vegetative period or drought resisting capacity can be cultivated. The dry farming is being encouraged by state officials to meet the requirements, but that too, will not harbour fruitful results in a long run, unless the soils are regenerated with their nutrients, otherwise many environmental problems may crop in.

Indira Gandhi Canal is making its expanding feet in the development of the area but the consequences are yet to see in future to come. It would be too early to say anything about it, but let us hope for the better. Two suggestions may be put forward with regards irrigation systems in arid zones :

1. In 1980s, some West German scientists engaged in irrigation mechanism and techniques in Arid lands, launched experiments in Egyptian arid lands and later shifted to Arabian Countries. They suggested that in arid and semi-arid lands 'Trench' irrigation farming is the best techniques suited to these areas. These trenches may be provided with bed sheets under-neath the depth of 2 meters which prevents or check the water percolation.

2. Secondly, there is a constant problem of saline water in arid lands to which water purifying process is needed. A Greenhouse process for salt water treatment has been proved to be a useful device in many countries of the middle East, which is a little bit different from the brackish water device. Though, Central Arid Zone Research Institute, (CAZRI) Jodhpur, also has developed a salt water processing plant but, has yet to be commissioned on a practical purposes.

The eastern region quite in contrast to the west, enjoys far better conditions of soil moisture and fertile soils of Trans-Yamuna plain. Bharatpur, Alwar, Jaipur and parts of Tonk and Swaimadhopur districts are constituted with alluvial to loamy soils which are rich in humas contents and other nourishing elements form a fertile tract. The moisture conditions are very favourable for the growth of plants which are available for the longer period of the cropping season, making it a better off region with number of crops. But farmers are caught with by the shooting market prices of the commodity grown in fields and therefore, they cultivate only those crops which may fetch them a good profit with less inputs. A case of mustard cultivation in Alwar and Bharatpur districts may well be exemplified in vicinity of the competing market prices of mustard against wheat crop. This factor has changed the whole complexion of the existing cropping pattern of wheat to mustard and other oil-crops. Mustard needed less water and the inputs as compared to wheat or any other crop of the Rabi season. But this pattern too is being overlapped by millets as one proceeds forward. In the changing cropping patterns the semi-arid parts of Jaipur region gives the clear cut impression of the uter fringe of the desert where aridity increases gradually towards west. The peripheral districts are well marked along with the line of aridity limit running from Jhunjhunu, Churu, Sikkar, and Jaipur districts.

Much of the problems related to the methods and techniques of irrigation causing concern to the irrigation engineering. Because, it relates the place and the prevalent system. The new system of irrigation technique like, Basin irrigation, Trench irrigation and sprinkling water irrigation, differ in areas of topography and terrain. Moreover it depends on the advancement of the farming technology. Thus mechanisation of farm-irrigation is a problem of all the developing countries especially those who are engaged in a farm-economy. There cannot be better co-ordination between new span of irrigation techniques and the existing old farm technology. If it is introduced on unmanaged farm it will not yield the desired results. Rudimentary farm systems cannot be matched with the new techniques.

Irrigation whether day or night was a matter of choice till the introduction of 'Fery' (on turn) system. Now it has become a business in the hands of the gate-operators and who control the release of water at the canal sluice. Hence it is a question of ferry or turn at which farmers have to wait till their number comes, and thus day and night irrigation system has lost its effectiveness and the significance also.

It is well borne fact that early hours of the morning irrigation is generally good for some of the crops. Though, it is a complex matter and not necessarily be regarded true for all the crops even then, these are the crucial hours which makes the matter more important. In most of the Indian Villages even to day, farmers prefer to irrigate crops in the early hours rather than late, say at about 4. O'clock in the winter morning when the temperature is at the lowest, effect crops more due to ample of water and temperature in the soil regime. At this moment if water is supplied to the crops, it is observed that they regain their vitality and generate new power over coming the severity of winter and this sustain life in crops. In most of the rabi crops, irrigation in such critical hours, helps as a life-serving dose. Wheat, Gram, Lucerne are the crops directly affected by water.

Thus, it is very difficult to arrive at any exact conclusion regarding the water and crop relationships, as both are complex phenomena, changes their courses according to the environments and some times adapt themselves in the existing conditions or die. To provide water in time is to accelerate the process of growth otherwise the whole mechanism is reversed and sudden death is occurred. Thus, the use or misuse of water is reflected in the plant's stages of development. The judicious use of water lies in the fact that we get the better yield from plants/crops, which is our ultimate aim of the present study.

References

Alberto Gasser & Helma Zeb-Gasser; 1983; 'Nitrogen losses from the Soil; plant research and development. Vol. 17, Institute of Scientific Cooperation, W.Germany, pp. 61-67.

Bhalla, L.R.; 1984; Rajasthan Ka Bhugol, Kuldeep Prak., Ajmer.

Critchfield, H.J.; 1975; General Climatology, Climate and Agriculture, Printice Hall, pp. 265-301.

De Martonne, E.; 1926; Une pouvelle function climatologique; L' indice d' aridite Meteorologie, 2,pp.449-459.

Dettmann, K.; 1981; 'Agriculture in the canal irrigation Region of Pakistan, Punjab- Applied geography & Development. Vol. 18, p. 111 -Berlin.

Directorate of Eco.& Statistics, Govt. of Raj.; 1982 ; Basic Statistics of Rajasthan, Jaipur.

Gdussen, H.; 1955;Expression des milieux per de formules ecologiques; Leur representation cartographie, in colloques mt. natn. Rech. Scient. 59 (1955) , pp. 257-269.

Jatzold, R.; 1976; Isolines of humid months as an aid to agricultural planning taking Kenya as an example' Applied sciences and development, 16, pp.330-350.

Jatzold, R.; 1977; Humid month Isolines as an aid in agricultural planning, Tubingen Applied sciences and development, 9, pp. 140-158.

Koppen, W.; 1918; Classification of climate according to temperature, precipitation and annual cycle, petermanns Mitt. 64, pp. 193-203.

Kumar, M.;1977; Estimation of ground water resources of western Rajasthan. The natural resources of Rajasthan, Vol.2, Jodhpur, pp. 591-599.

Mishra, P.R.& Handa D.P.; 1977; Prospect of Harnessing Rain Water to maximum benefits in Eastern Rajasthan. The Natural Resources of Rajasthan, Vol. 2, Jodhpur, pp. 629-637.

Mishra, V.C.; 1967; Geography of Rajasthan, NBT, New Delhi.

Mosonyi, E., Linder, A., Francke, F.; 1984; Modaling Techniques in the Planning of Irrigation Systems, Applied Geog. & Development, Vol. 23, pp. 94-108, Berlin.

Penck, A.; 1910; An attempt at climate classification based on physiogeographical factors; Berlin, Sitz-Ber., Kgl. Preub. A Kad. Wiss., pp. 236-246.

Penman, H.L.; 1948; Natural evaporation from open water, Base soils and grass, Proc. Roy. SOC(A). p. 193.

Peter Wolff; 1983; 'Technical aspects of mechanisation of irrigation farming; Applied Geography and Development, Vol. 19, W.Germany, pp.1-7.

Rehm,S.; 1982; 'Competition between Animal and Plant Production in World Nutrition', Plant Research & Development, Vol.15, pp.117, Berlin.

Roonwal, M.L.(Ed); 1977; The Natural Resources of Rajasthan, Vol. I & II, University of Jodhpur, Jodhpur.

Schreiber, D.; 1981; A Comparative Examination of Threshold values for Dry Mouth Taking the Situation in Kenya as an example, Applied Geography and Development,17, Tubingen, Berlin,pp. 29-43.

Schiedeeken, W.; 1981; Humidity and Cultivated Plants, Applied Geography and Development, 17, Tubingen, Berlin, pp. 45-57.

Singh, S.D.& Mann, H.S.; 1979; Optimization of Water Use and Crop Production in an Arid Region, CAZRI, Jodhpur.

Thornthwaite, C.W. & Mathur, J.R.; 1957; Instructions and Tables for Computing Potential Evapo Tran-piration and the Water Balance, Pub. in Clim., 10, 3.

Walker, H.H.; 1984; "Problems of Organisation of Irrigation Projects...." Applied Geog. & Development, Vol, 23, pp. 20-45, Berlin.

Werner Doppler; 1983; 'Irrigation as basis for development'. Plant Research and Development . Vol. 17, Institute for Scientific Cooperation, W. Germany, pp.61-67.

Index